DATE			

Cyril Burt

Cyril Burt

Fraud or Framed?

Edited by

N. J. MACKINTOSH

Department of Experimental Psychology
University of Cambridge

OXFORD NEW YORK TOKYO
OXFORD UNIVERSITY PRESS
1995

Oxford University Press, Walton Street, Oxford OX2 6DP
Oxford New York
Athens Auckland Bangkok Bombay
Calcutta Cape Town Dar es Salaam Delhi
Florence Hong Kong Istanbul Karachi
Kuala Lumpur Madras Madrid Melbourne
Mexico City Nairobi Paris Singapore
Taipei Tokyo Toronto
and associated companies in
Berlin Ibadan

Oxford is a trade mark of Oxford University Press

Published in the United States
by Oxford University Press Inc., New York

© Oxford University Press, 1995

A catalogue record for this book is available from the British Library

Library of Congress Cataloging in Publication Data
Cyril Burt : fraud or framed? / edited by N.J. Mackintosh.
Includes bibliographical references and index.
1. Burt, Cyril Lodowic, Sir, 1883–1971. 2. Fraud in science—
England—Case studies. 3. Psychologists—England. I. Mackintosh,
N. J. (Nicholas John), 1935– .
BF109.B88C87 1995   150'.92—dc20   95-93
ISBN 0 19 852336 X

Typeset by Downdell, Oxford
Printed and bound in Great Britain by
Biddles Ltd, Guildford and King's Lynn

Preface

CYRIL BURT DIED in 1971, full of honours and after a career spanning more than 60 years. Appointed to the chair of psychology at University College London in 1932, he retired in 1950, but remained active for the next 20 years, virtually until the day he died. He was knighted in 1946, one of only three British psychologists to have been so honoured, and was widely regarded as the most significant and influential educational psychologist of his generation, whose research on educational attainments, juvenile delinquency, intelligence testing, and factor analysis were landmarks in their fields. But, as Arthur Jensen notes in Chapter 1, Burt's posthumous notoriety has comfortably surpassed the fame he achieved in his lifetime. Within a year of his death, Leon Kamin of Princeton University was giving a series of lectures in the US in which he poured scorn on Burt's research on the genetics of intelligence. By 1976, Burt was being explicitly accused of fabricating data to prove that intelligence was inherited, and although these accusations were vigorously disputed, his fate seemed sealed with the publication in 1979 of his official biography by Leslie Hearnshaw, a widely respected historian of British psychology. With access to Burt's unpublished correspondence and diaries, Hearnshaw reluctantly concluded that the charges of fraud were justified—and added a couple more of his own for good measure.

Hearnshaw's biography received almost uniformly favourable reviews, and his conclusions were accepted without further ado by the British Psychological Society, which formally concluded that Burt had been guilty of fabrication of data. And there the matter might have rested. Most of Burt's earlier defenders either acknowledged their error, or remained prudently silent. But two books published some 10 years later, one by Robert Joynson, the other by Ronald Fletcher, succeeded in reopening the case. Working independently, both concluded that the charges against Burt had not been proved and that, at the very least, he deserved the benefit of the doubt. Although neither book was as favourably received as Hearnshaw's, many reviewers accepted that they had raised legitimate doubts.

Where does the truth lie? Was Burt the innocent victim of politically motivated calumny? Have Joynson and Fletcher glossed over some of the evidence against him? It is almost certainly too late to answer many of the questions raised by the Burt affair with any certainty. Burt can neither defend himself against the charges brought against him nor answer a host of quite legitimate questions. Many other principal actors in the affair are also dead. But if much uncertainty will inevitably remain, it does not follow that there can be no grounds for deciding whether Hearnshaw or Joynson and Fletcher are nearer the truth. It may still be possible, by a close examination of the published record, to narrow down the area of uncertainty—to show, for example, that Hearnshaw was probably wrong on some counts, but that Joynson and Fletcher may have been wrong on others. What I, and the other contributors to this book, have tried to do is to examine the available evidence as carefully and dispassionately as possible, to distinguish between those charges or counter-charges which are probably false, those which may well be true, and those where we shall probably never know. We have tried, as best we can, to eschew exaggeration, rhetoric, and calumny. Of course, Burt's friends and admirers were incensed by what they saw as a mixture of jealous rancour, revenge, and political propaganda. Equally, some of his critics were outraged by what they saw as the casual and cavalier way in which Burt presented much of his data and the reverence with which those data were accepted—an acceptance which they too saw as politically motivated. Academic writers also have their own, professional axes to grind: the thrill of the intellectual chase and the pleasure of scoring points in debate can often override a nice regard for accuracy or an impartial willingness to acknowledge evidence that does not suit one's case. How far my fellow contributors and I have succeeded in avoiding these pitfalls, readers must judge for themselves. Our intention at least has been to avoid, as far as possible, simply refighting old battles and, as Chapter 7 should make clear, to distinguish between the question of whether Burt's data are fraudulent and the current status of the scientific issues they addressed.

Chapter 1, by Arthur Jensen, provides a brief outline of the major aspects of the Burt affair and thus serves as an introduction to what follows. Chapters 2–5 take up, in no particular order, the four main charges of fraud identified by Hearnshaw. In Chapter 2, Steven Blinkhorn discusses the history of factor analysis which, according to Hearnshaw, Burt falsified—at Spearman's expense and to his own advantage. In Chapter 3, I consider Burt's data on kinship correlations for IQ, in particular his sample of 53 pairs of separated identical twins. Chapter 4, by Nicholas Mascie-Taylor, examines a paper published by Burt in 1961 on intelligence and social mobility, and in Chapter 5 I discuss a paper published in 1969 in which Burt produced evidence of an apparent decline in educational standards over the 50 years since the outbreak of the First World War.

Finally, in Chapter 6, Hans Eysenck, who knew Burt well, recounts some of his recollections of Burt's behaviour from the 1930s to the 1950s, since one of the charges against Burt, which Hearnshaw accepted, was that he sometimes behaved dishonestly in his dealings with colleagues. Eysenck also discusses other cases in which eminent scientists have 'improved' their data in order to support their theories. And in a concluding chapter, I address the question of the scientific validity and impact of Burt's work.

I do not for a moment suppose that we have written the last word on Burt, or that we shall convince all those who have taken up public positions on the question of his guilt or innocence. I should like to believe that we have clarified some of the issues, and provided enough information for readers to form their own judgment.

Finally, I should like to thank my fellow contributors for their co-operation and cheerful acceptance of my editorial suggestions. I am also grateful to many other people, too numerous to mention, with whom I have discussed some of the issues raised in this book. I should particularly like to mention Dr Clare Burstall, formerly Director of the National Foundation for Educational Research, and Professor Hannah Steinberg, now at Middlesex University, who were kind and patient enough to answer my questions, even though they knew that they might not agree with my conclusions. And I must also thank Felicity Fildes, who retyped numerous drafts of the manuscript and tracked down an even larger number of obscure references.

Cambridge   N.J.M.
December 1994

Contents

Contributors

S. F. BLINKHORN
Psychometric Research & Development Ltd., Brewmaster House, The Maltings, St Albans, Hertfordshire, AL1 3HT

H. J. EYSENCK
Institute of Psychiatry, De Crespigny Park, Denmark Hill, London, SE5 8AF

A. R. JENSEN
Educational Psychology, University of California, Berkeley, 4511 Tolman Hall, Berkeley, California, 94720, USA

N. J. MACKINTOSH
Department of Experimental Psychology, University of Cambridge, Downing Street, Cambridge, CB2 3EB

C. G. N. MASCIE-TAYLOR
Department of Biological Anthropology, University of Cambridge, Downing Street, Cambridge, CB2 3DZ

ONE

IQ and science: the mysterious Burt affair*

A. R. JENSEN

THE CASE OF Sir Cyril Burt is probably the most bizarre episode in the entire history of academic psychology. This can be attributed to a combination of elements—the controversial subject of Burt's major research, his unusual personality, his widely acknowledged accomplishments, and the damaging accusations levelled against him after his death. Indeed, Burt's posthumous notoriety exceeds even the considerable fame he enjoyed during his long career.

In his famous study of the IQs of fifty-three pairs of MZa twins—monozygotic (identical) twins reared apart—Burt had shown a high correlation (0.77) between the general intelligence of separately raised twins. What became known as the 'Burt scandal' surfaced in 1976, five years after Burt's death. He was accused of faking data and fabricating both research assistants and co-authors to lend his deception authenticity. The main thrust of the attackers' effort was to discredit Burt's major theory—that genetic factors are strongly involved in human intelligence—as well as the body of research that supports it. Still, Burt was not without his defenders. A number of scholars, mainly former associates, rose to his defence, writing articles and letters to newspapers and making television appearances. The controversy continued for three years.

Then, in 1979, Burt's guilt seemed virtually clinched when Britain's most highly respected historian of psychology, Leslie Hearnshaw, published *Cyril Burt, psychologist*, which appeared to be a carefully researched and impartial biography of Burt. Hearnshaw had exclusive access to Burt's private correspondence and diaries, which no one else had yet seen. Thus, the biography was almost universally accepted as the last word on the subject and even persuaded most of Burt's supporters. The devastation of Burt's once exalted reputation was a gleeful triumph to his detractors and

* An earlier version of this chapter appeared in *The Public Interest*, **105**, Fall 1991, 93–106, published by National Affairs, Inc. It is reprinted, in revised form, with the permission of the publishers.

a tragedy to his admirers. With sighs of relief all around, the matter appeared to be settled at last. Or so most of us thought.

Recently, the investigative efforts of two British scholars, psychologist Robert B. Joynson and sociologist Ronald Fletcher, have reopened the case. Neither man knew Burt personally or ever had any previous connection with Burt's research or the 'IQ controversy'. The two investigators, working entirely independently, devoted several years to carrying out what appears to be painstaking detective work on the Burt affair. Both Joynson's 1989 book *The Burt affair*, and Fletcher's 1991 work *Science, ideology, and the media: the Cyril Burt scandal*, critically question every accusation and meticulously sift through the evidence. Although their accounts differ markedly in organization and style, with regard to the main charges against Burt the two authors reach the same conclusion: *not proven*.

A brilliant eccentric

Before getting into the details of this perplexing case, it is important to know just who Burt was, personally and professionally. Certain features of Burt's personality, and especially his area of research, helped the scandal sprout and flourish.

Sir Cyril Lodovic Burt (1883–1971) was long regarded as a towering figure in the history of British psychology. The first British psychologist to be knighted (a distinction bestowed on only two other psychologists to date), Burt was renowned for his intellectual brilliance and scholarly industry. After graduating from Oxford University, where he studied classics, mathematics, physiology, and psychology, Burt worked for four years as an assistant to the celebrated neurophysiologist Sir Charles Sherrington at Liverpool University. In 1913, he was appointed to the post of Psychologist for the London County Council, initially combining this with the job of part-time assistant lecturer in experimental psychology at Cambridge. His London job put Burt in charge of psychological research and applied psychology, including the development of mental and scholastic tests, for the entire London school system. In this setting he became one of the world's leading educational psychologists and psychometricians, developing new tests, conducting surveys, and founding child guidance clinics and a special school for the handicapped. Burt conducted pioneering research on juvenile delinquency and mental retardation. He reported some of these studies in beautifully written books that became classics in their field: *The young delinquent* (1925), *The subnormal mind* (1935), and *The backward child* (1937).

During much of the period that Burt held his appointment with the London County Council, he also occupied the chair in educational psychology at the University of London. In 1932 when Charles Spearman, one of the great pioneers of mental testing, retired as head of the Department of Psychology at University College London, Burt was appointed to his position, at that time the most influential in British psychology.

Burt retired from that post in 1950, at the age of sixty-eight. The last twenty years of his life were spent in a rather reclusive manner, living in a large London flat with a secretary–housekeeper, editing journals, and writing books and articles. He was remarkably prolific even in his old age. Following his retirement he published over 200 articles and reviews. And those were only the items published under his own name. In addition, as his most notable eccentricity, he wrote a considerable number of articles, mostly book reviews, under various pseudonyms or initials of unidentifiable names. He worked steadily almost until the day he died, at the age of eighty-eight.

Two areas of research for which Burt was best known were factor analysis and the genetics of intelligence, fields in which his mathematical aptitude could be used to great advantage. In both fields, Burt's work was ground-breaking. He expertly adapted new developments in quantitative genetics to the study of human behavioural traits. Kinship correlations are the essential data for quantitative genetic analysis. Beginning quite early in his career, while still working in the London schools, Burt started collecting IQ and scholastic-achievement scores on twins and various other related groups. Between the years 1943 and 1966 he published many theoretical and empirical studies dealing with the inheritance of intelligence; his last paper on this topic was published posthumously in 1972.

It was particularly this genetic aspect of Burt's psychometric studies of individual differences that seemed to have such controversial educational and social implications. Egalitarian intellectuals tended to view the so-called 'nature–nurture' question as a political issue rather than a scientific one, and so the potential controversy extended to a much larger arena than just the field of behavioural genetics. Burt himself, however, seldom expressed any interest in politics and never joined any political party. His knighthood was awarded by a Labour government.

Burt's personality is a more puzzling matter. I knew Burt personally and enjoyed numerous visits with him in the last two years of his life. But it was obvious to Burt that I was an admirer, and probably his relationship to me, always friendly and generous, was not entirely typical of his dealings with his academic colleagues or students. As is clear from Eysenck's chapter below, opinions of Burt vary widely among this group, ranging from the highest esteem to bitter denigration, at times both coming from the same observer. There are only three characteristics about which one finds complete agreement: Burt's exceptional intellectual brilliance, his extraordinary general erudition, and his untiring industry.

The less favourable impressions of Burt registered by a few of his former students, colleagues, and acquaintances mention his egocentrism and personal vanity, his autocratic manner in running his department, his insistence on getting his own way, and his obsessive need to have the last word in any argument. Also, as a noted colleague Philip E. Vernon wrote, 'It seemed difficult for him to allow his past students or followers to branch out and publish contributions which went beyond his views.' Added Vernon, 'Although Burt gave immense amounts of help to students and others, he could not brook any opposition to his views, and often showed paranoiac tendencies in his relations with colleagues and critics.'

Disappearing data

Perhaps the only means by which to evaluate Burt objectively is to judge him by the published work he left behind. His strictly theoretical work on factor analysis and on the polygenic theory of intelligence was an important contribution to the field; it provided a heuristic methodology for the study of individual differences that many other researchers applied in subsequent studies. The reliability of Burt's empirical research, by contrast, has frequently been called into question.

A few days after the news of Burt's death in 1971, I wrote to Miss Gretl Archer, who was Burt's private secretary for over twenty years, to request that she preserve the two or three tea crates of old raw data that Burt had once told me he still possessed. I told Miss Archer that I would travel to London the following summer to go through this material. I supposed it included IQ test data on twins, in which I had an interest and which I thought could be used in certain newer kinds of genetic analysis that Burt had not attempted. Miss Archer replied that all of these data had been destroyed within days after Burt's death, on the advice of Dr Liam Hudson, professor of educational psychology at Edinburgh University. He had come to Burt's home soon after the announcement of Burt's death. Miss Archer, distraught and anxious to vacate Burt's large and expensive flat in Hampstead, had already arranged for the disposition of Burt's library and correspondence files (which were turned over to his biographer Hearnshaw), but expressed concern to Hudson about what to do with these boxes of old data. Hudson looked over their contents and advised that she burn them, as being no longer of any value. Miss Archer said she believed the boxes included the data on twins, and later expressed regret that she had acted on Hudson's advice. The account I received from Miss Archer of this event was corroborated by Hudson himself, in an interview with *Science* staff writer Nicholas Wade (1976). Hudson explained that he thought Burt's old data sheets were probably unintelligible to anyone but Burt himself.

Accusations of fraud

The first public accusation of outright fraud appeared on 24 October 1976 in the London *Sunday Times*, under the striking headline: 'Crucial data was faked by eminent psychologist', written by Oliver Gillie, the *Sunday Times*'s medical correspondent. Within days the story was repeated in the mass media around the world. Gillie followed with other sensational articles under headlines such as 'The great IQ fraud' and 'The scandal and the cover-up', writing of 'outright fraud', and calling Burt a 'plagiarist of long standing'.

These charges were not based on anything new involving Burt's data, several peculiarities in which had already been pointed out two years earlier by myself (Jensen 1974) and by Kamin (1974). Gillie's allegations rested largely on the claim that he had been unable either to locate in person or to find any trace of two women—Margaret Howard and J. Conway—who were credited with assisting Burt in his research on twins. Howard was a co-author of one of Burt's most important articles on twins and Conway was named as the sole author of an article that was actually written by Burt himself, according to his secretary. These two women could not be traced or even identified with certainty by anyone available for questioning who had been associated with Burt. The 'missing ladies', as Gillie called them, gave him licence to claim that Burt's data were, as he put it, 'faked'.

Gillie credited Professor Jack Tizard (since deceased, but then a psychologist at London University's Institute of Education) with helping him search for the 'missing ladies'. Tizard, although he had scarcely known Burt personally, became an active participant in the attack, giving Gillie information and advice on how to go about it. I was well acquainted with Tizard, having spent two years at London University's Psychology Department, where he also taught at the time. He was, as his wife explained to Joynson, a 'passionate egalitarian'. Tizard was also quite outspokenly anti-hereditarian and anti-Burtian.

The day after Gillie's sensational charges of fraud in *The Sunday Times*, there appeared in *The Times* an interview with Tizard, titled 'Theories of IQ pioneer "completely discredited"'. It began: 'The theory of Sir Cyril Burt . . . that man's intelligence is largely caused by heredity was now completely discredited, Professor Jack Tizard, Professor of Child Development at London University, said yesterday . . . Professor Tizard said the discrediting of Burt's work cast doubt on his whole line of inquiry.'

It seems highly likely that the main steam behind the attack on Burt may have been the fervent wish of environmentalists such as Tizard to discredit the theory of polygenic inheritance of mental ability and other behavioural traits of obvious personal, educational, and social importance. Such indeed was the leitmotif in the popular press and on television, both in England and America. (It even predominates in accounts of Burt in some psychology textbooks.) Since ideological propaganda depends not on facts, but on images, impressions, and prejudices, the anti-Burt campaign naturally avoided the fact that Burt's research was in line with the consensus of other expert studies on the heritability of IQ.

Hearnshaw's biography

When the scandal broke in the media it was already known in psychological circles that Professor Hearnshaw had been working for several years on what would become the 'official' biography of Burt. Because of Hearnshaw's well-recognized scholarly credentials as a historian of psychology, and the fact that he had no prior involvement in the IQ controversy or in any other aspect of Burt's activity, his objectivity and credibility in the Burt case were unblemished. Also, he had delivered a beautiful eulogy at Burt's memorial service and was commissioned to write the biography by Burt's sister, who made available Burt's diaries and correspondence. It was everyone's reasonable expectation that Hearnshaw's forthcoming biography of Burt would become generally regarded as the last word on the matter.

Although Hearnshaw was already in the late stages of his writing, it was of course mandatory that his biography deal fully with the scandal. Several of Burt's detractors took this opportunity and made further accusations that had not previously come to light. Among these were Ann and Alan Clarke, two psychologists at the University of Hull, who had obtained their PhDs under Burt in 1950, but who, in a textbook published in 1974, had noted certain 'puzzling features' and 'suspiciously perfect' data in some of Burt's published work.

Burt's detractors were obviously successful in convincing Hearnshaw of Burt's guilt. When his massive and well-written biography was published in 1979, his conclusions of guilt on several counts were widely accepted, even by most of Burt's former defenders. The Council of the British Psychological Society (BPS) endorsed Hearnshaw's conclusions and officially declared Burt's guilt in a 1980 booklet entitled *A balance sheet on Burt* (Beloff 1980). The 'balance sheet', however, was anything but balanced. Both Tizard and Alan Clarke were members of the BPS Council when it planned its official pronouncement on Burt. Among the seven presenters in the *Balance sheet* were Hearnshaw, Gillie, Ann Clarke, and Alan Clarke. As expected, they all roundly condemned Burt, while the remaining three contributors, who had never visibly done any research on the Burt affair themselves, acquiesced in the official pronouncement and wrote only in general terms on research methodology and scientific fraud. As far as I know, there was no attempt to question any of the evidence used to support the various charges against Burt.

Why were so many convinced by Hearnshaw's book? I had reviewed the manuscript for the publisher and praised it highly. Its cool-headed, judicious style evinced none of the rancour or anti-hereditarian rhetoric typical of many of Burt's detractors. What seemed to be the crucial evidence in Hearnshaw's exclusive possession were Burt's diaries and correspondence. The diaries covered the much examined period in Burt's career (1953–60) during which he published articles suggesting that he had acquired data on new sets of twins. Hearnshaw gives the impression that the diaries were quite complete and detailed, recording even such insignificant things as Burt's having tea with a friend, taking a walk, or getting a haircut. Surely one would think that anything as important as locating and testing newly discovered sets of MZa twins would be mentioned in the diary, if this had actually occurred. Their complete absence in the diaries would seem to be damning evidence.

However, when the diaries are closely examined, as they were by Joynson and by Fletcher (whose book also reproduces all the entries in Burt's diary for one month), this negative evidence of not having collected any new sets of twins (at least after 1953) suddenly becomes unimpressive. The reason is that Burt's diaries seem to record *nothing* but utter trivia; for example, there is no mention at all of the death of Burt's personal secretary of many years or of Burt's attending her funeral, which other records show he did. The diaries read more like a simple date book, with the briefest possible notations. Furthermore, some 55 per cent of all the dates during the whole period covered by the diaries show no entries at all, and there are a number of periods of several consecutive months without a single entry. So the absence of entries on MZa twins (or other kinship data) in the diaries, and the lack of any mention of his former assistants, Howard and Conway, become unconvincing evidence for the charge that Burt faked his data. Yet Hearnshaw's rather misleading report of the nature of these diaries had finally convinced almost everyone that Burt had committed fraud.

The nearest thing to a 'smoking gun' in Burt's diaries is the single entry, 'calculating data on twins for Jencks'. This item does give the reader pause. In 1968, Christopher Jencks, a Harvard sociologist, had requested from Burt a listing of the IQs and socio-economic ratings of each of the fifty-three MZa twin pairs on which the correlations were based in an important article Burt published in 1966. The crucial question here is does 'calculating data' mean deliberately *concocting* data to fit the already published correlations and other statistics? Or could it mean something else, perhaps just assembling data from various other tables or test sheets, or matching up the socio-economic information on the subjects from separate data files? No one really knows. There is indisputable evidence from Burt's correspondence that he told 'white lies' to Jencks and other correspondents about the reasons for his delayed replies to their inquiries (such as claiming to have been out of town), but this can hardly be construed as evidence that he fabricated the MZa data he sent to them.

Another source of suspicion is the fact that Burt wrote in 1971 to Sandra Scarr, a noted behavioural geneticist then at the University of Minnesota, in reply to her request for a copy of his data on fifty-three sets of MZa twins. In his letter he also gave the IQ scores and details on three new sets of MZa twins (Scarr had sent me a copy of this letter, which I passed on to Hearnshaw). I was especially puzzled by this, because about two months after Burt wrote that letter, I had personally discussed twin research with him, and had even mentioned the possibility of looking for more sets of MZa twins in London. Yet he never mentioned having found the three new sets of twins he had described to Scarr. It seems improbable to me to attribute Burt's silence on this point to a lapse of memory. Although he was then eighty-eight years old, his memory was phenomenal for a great many other things, including the technical details of one of my own studies that I had described in conversations with him two weeks previously. Again, however, Burt's failure to mention the three new sets of twins to anyone but Scarr remains negative evidence—apparently damning, but still inconclusive.

The case for the defence

It is impossible in this brief account to do justice either to the great wealth of detail in Hearnshaw's biography or to the extensive and fine-grained investigation presented by Burt's defenders, Joynson and Fletcher. Consequently, the case for the defence can only be characterized in the most general terms. The line of defence argued by Joynson and Fletcher consists of two main tactics: (1) they show the previously unsuspected flimsiness, misrepresentation, and even in some cases factual non-existence, of the supposedly damning evidence; (2) they closely examine the points that had aroused suspicion and provide alternative innocent explanations that seem at least as plausible as the 'guilty' explanations promoted by Burt's accusers. Four examples are given below.

Point one. Burt's assistants Howard and Conway could not be found, nor could their existence at any time be definitely established.

Counterpoint. Howard and Conway presumably worked for Burt only prior to the Second World War and, assuming they were still alive when sought in 1976, they would have been quite elderly. Burt's secretary informed me that Burt had told her that Conway had emigrated, perhaps to Australia. Other persons that Burt mentioned in his articles and who at first were also suspected of being fictitious were later identified. Burt's articles were not explicit about exactly when Howard and Conway actually collected the data on twins, and he was perhaps deceptive in leaving the impression that they were still giving IQ tests to twins even after 1955. My own hunch is that his personal vanity made him want to appear more actively engaged in ongoing research in his old age than he actually was, and so he obscured the 'when and how' of his data collection, an implicit deception that later engendered doubts about the data's authenticity.

Point two. Neither Burt's diaries nor correspondence provide evidence that Burt or any identifiable former assistants tested any new sets of MZa twins after Burt officially retired in 1950. Yet he added new twin data to his studies published in 1955 and again in 1966.

Counterpoint. Virtually all of Burt's data were collected before the Second World War. After the first blitzkreig on London, University College had to be rapidly evacuated. All of Burt's data were hastily thrown into various boxes and stored in the basement; his department was moved to Wales for the duration. In a later bombing raid, the College suffered a direct hit. One of Burt's longtime associates, Charlotte Banks, testified that the twin data were retrieved piecemeal after the war. They were found in different boxes and at different times. Some had been misplaced and turned up only much later. Although Burt's articles implicitly made it appear that he was collecting new data, the fact was that he only analysed and reported for the first time old data that had been collected many years before. Burt's odd furtiveness in this regard justifiably undermined his posthumous reputation; regardless of whether or not one accepts Joynson and Fletcher's explanation about misplaced data, one must conclude that Burt's deception is inexcusable for a scientist.

These two points are examined below in Chapter 3.

Point three. In a 1978 feature article in *Science*, an American psychologist, Donald Dorfman, attempted to demonstrate statistically the fraudulent nature of the data from one of Burt's articles on social mobility and IQ, which showed results consistent with the hypothesis that the average social-class differences in IQ reflect genetic differences. Dorfman argued that the tables in the article fit the normal curve so closely as to be almost certainly faked. In other words, it was improbable that random subject samples would show the high degree of regularity seen in Burt's tables.

Counterpoint. Apparently Dorfman's haste to prove Burt a fraud precluded his reading Burt's article carefully. Burt explicitly indicated that he normalized the data in his tables. Two professors of mathematical statistics, at Harvard and the University of Chicago, strongly disputed Dorfman's argument. They pointed out that Burt's procedure of normalizing the frequencies, or fixing the marginal totals, was a statistically acceptable and not uncommon practice for this type of analysis. Jointly, they further stated that 'using Dorfman's inappropriate statistical techniques to detect fraudulent data would be to condemn a major portion, if not all, of empirical science as fabrication.' This issue is discussed at greater length in Chapter 4 by Mascie-Taylor.

Point four. Hearnshaw accused Burt of falsifying the history of factor analysis, belittling Charles Spearman's claims as the inventor of this technique, assigning major credit to Karl Pearson, the 'father of mathematical statistics', and aggrandizing his own contribution to the development of factor analysis.

Counterpoint. Actually, Burt's account of the history of factor analysis is correct, and Hearnshaw's verdict on this score is simply mistaken.

Pearson, in 1901, invented what today is known as principal axes or principal components analysis, although Pearson did not apply it to psychological data. But this technique was, and still is, widely used in psychological research, and it closely resembles virtually all other present-day methods of factor analysis. In contrast, Spearman's original method of factor analysis has been obsolete for more than fifty years and is seldom explicated in modern textbooks of factor analysis. Invented independently of Pearson's contribution a few years later, Spearman's formulae are no longer used because they can extract only a single factor (a general factor, or g) from a correlation matrix and the method is correctly applicable only to a limited class of matrices (hierarchical matrices with a rank of unity).* Burt's contribution occurred later, with the invention of a method of multiple factor analysis known as 'simple summation'. This method is similar to the 'centroid' method later developed by Thurstone. In the days of mechanical calculators, both Burt's and Thurstone's methods had the advantage of being less laborious to compute than Pearson's principal axes. Hence, they were widely used for many years until the advent of electronic computers made mathematically more elegant and exact procedures practicable. Chapter 2, by Blinkhorn, provides a more detailed account of these issues.

A final point

Interestingly, Hearnshaw does not address in detail one of the most serious charges commonly levelled against Burt—that his calculations themselves were fraudulent. My own 1974 examination of Burt's studies did find a number of peculiarities in his data. Almost all of the errors, however, amounted to the kind of careless copying or proofreading mistakes one could reasonably attribute to Burt's advanced age—he wrote most of the articles in question after the age of seventy-five. Moreover, as Joynson notes in *The Burt affair*, a number of the apparently erroneous twin

* The clearest discussion of the limitations of Spearman's method of factor analysis that I have found in the literature is by Thurstone (1947, Chapter XII, especially pp. 279–81). He states (p. 268) that the method is applicable only to a matrix of unit rank (i.e. a matrix with only a single-common-factor when communalities are in the diagonal) and also that, after solving for the first factor loadings using Spearman's single-factor formulae, attempts to extract additional factors in the same manner from the residuals will yield theoretically incorrect solutions; he presents a mathematical proof of this conclusion (p. 280). He notes that the application of the single-factor formulae to a correlation matrix can be justified only by regarding the result as a single-factor description of the correlation matrix. In that case the first-factor residuals are regarded merely as variable errors which, if the matrix were not of unit rank, would be too large to be acceptable by Spearman's criterion of 'vanishing tetrads'. The method is obviously stymied in the face of a matrix of correlations that reflect multiple factors. In practice, Spearman always began his analysis by using his vanishing tetrads criterion for discarding any variables in the correlation matrix that broke its hierarchical pattern, or unit rank, before applying his formulae for calculating the variables' loadings on the single, or general, factor in the matrix.

and sibling correlations in fact suggest a *decrease* in the heritability co-efficient, strengthening the argument for environmental causation of IQ differences. Burt would hardly engage in deliberate fraud in order to bolster a conclusion completely opposite to his own.

Conclusion

What is the moral of this curious story? A talented scientist who works largely alone makes a good many personal enemies. He is sometimes careless and eccentric in his presentation of his studies. He becomes a prominent public figure. Most important, he develops politically incorrect theories on socially sensitive topics. This combination of factors gives his opponents—aided by sympathetic journalists—ample ammunition to attack his reputation.

Such is the essence of the Burt affair. Joynson and Fletcher have dis-proved some of the accusations and suspicions levelled against Burt, but not all, and not completely. There is room left for doubt. Whether one gives the benefit of the doubt to Burt or to his detractors is still another matter. A convincing defence of Burt is handicapped by his undisputed personal eccentricities and petty foibles, as well as by his failings as an empirical scientist. Since it is next to impossible to prove a negative, no one can confidently proclaim Burt's complete innocence of all charges. He may be guilty of simple carelessness. But the burden of proof rests squarely on those who have proclaimed Burt guilty of fraud. Virtually all of their evid-ence has proven so flimsy that I believe an impartial jury would rule out the verdict of fraud, not just on the grounds of 'not proven', but simply as not plausible.

A final judgement on Burt would probably not much interest historians of psychology if it had turned out that his conclusions about the heritability of intelligence were wrong. But in the twenty years since Burt died, many scientifically rigorous studies—including a recent MZa-twin replication virtually identical to Burt's—have substantiated the theory that individual differences in intelligence are strongly conditioned by genetic factors. Experts in behavioural genetics now generally agree on this central point. As all the smoke and fumes of the Burt affair dissipate, this should be cause for optimism: the field of behavioural genetics appears increasingly ready—controversy notwithstanding—to behave as a science like any other.

References

Beloff, H. (1980). A balance sheet on Burt. *Bulletin of the British Psychological Society* (suppl.), **33**.
Burt, C. L. (1925). *The young delinquent*. University of London Press.

Burt, C. L. (1935). *The subnormal mind.* Oxford University Press, London.

Burt, C. L. (1937). *The backward child.* University of London Press.

Burt, C. L. (1966). The genetic determination of differences in intelligence: a study of monozygotic twins reared together and apart. *British Journal of Psychology,* **57,** 137–53.

Clarke, A. D. B. and Clarke, A. M. (1974). *Mental deficiency* (3rd edn). Methuen, London.

Dorfman, D. (1978). The Cyril Burt question: new findings. *Science,* **201,** 1177–86.

Fletcher, R. (1991). *Science, ideology, and the media: the Cyril Burt scandal.* Transaction Publishers, New Brunswick, NJ.

Gillie, O. (1976). Crucial data was faked by eminent psychologist. *The Sunday Times,* 24 October. London.

Hearnshaw, L. S. (1979). *Cyril Burt: psychologist.* Hodder & Stoughton, London.

Hearnshaw, L. S. (1980). A balance sheet on Burt. In *Bulletin of the British Psychological Society* (suppl.), **33** (ed. H. Beloff).

Jensen, A. R. (1974). Kinship correlations reported by Sir Cyril Burt. *Behavior Genetics,* **4,** 1–28.

Joynson, R. B. (1989). *The Burt affair.* Routledge, London.

Kamin, L. J. (1974). *The science & politics of IQ.* Erlbaum, Hillsdale, NJ.

Thurstone, L. L. (1947). *Multiple factor analysis.* Chicago University Press.

Wade, N. (1976). IQ and heredity: suspicion of fraud beclouds classic experiment. *Science,* **194,** 916–19.

TWO

Burt and the early history of factor analysis
S. F. BLINKHORN

AT THE TIME the Burt scandal originally broke in 1976—a day I remember as clearly as the day Kennedy was assassinated—I was doing some initial reading for my own PhD, which was concerned in a technical sense with certain developments in confirmatory factor analysis. By the time Hearnshaw's biography of Burt appeared in 1979, the PhD was awarded and I had moved on to other concerns. So it was not until *Nature* sent me Stephen J. Gould's book *The mismeasure of man* for review that I really took note of the fact that doubt had been cast on Burt's own account of his place in the early history of factor analysis. Gould's own book was so manifestly partisan—in my review I described it as 'researched in the service of a point of view rather than written from a fund of knowledge'—and so surprisingly coarse grained in its account of the origins of factor analysis and the factor-analytic school in psychology, that I thought little more of it at the time. It is too easy to expect non-specialists to be aware of and take into account in popular works the distinctions that are important to specialists, as is doubt-less the experience of most scientists watching a television documentary on a subject which touches their professional concerns.

A few years later, *Nature* sent another book for review, this time *The intelligence men* by Raymond Fancher. Here, to my astonishment, I read (p. 176):

With remarkable ease for a person totally lacking in formal mathematical training, Burt mastered the basic factor analytic techniques and soon began making original contributions of his own.

This was supposed to have happened when Burt took up his chair at University College London, more than 20 years after he published his first paper, and long after what, in recalling my initial reading of the early literature, had seemed to me to be his original contributions to the field.

Both Gould and Fancher repeat the accusation made by Hearnshaw (1979) that Burt falsified the history of factor analysis in his later work. Specifically, Hearnshaw proposed that while Spearman remained alive

Burt never departed from the universally acknowledged account that Spearman was the originator of factor analysis, but that once Spearman was dead, Burt himself laid claim to that distinction.

Finally in 1989, *Nature* sent me Robert Joynson's book *The Burt affair*, in which I read an account of Burt's early contributions to the development of factor analysis that chimed so well with my recollections of my reading more than a decade earlier that I went back to some original sources and checked what Joynson had to say against what appeared in the published record. Then, as now, I came to the conclusion that in this matter Joynson's account is substantially correct. But it is written in the context of a riposte to the general run of accusations against Burt, and does not fully take into account the broader context of the history of the development of the various techniques that have been called factor analysis.

The purpose of this chapter, therefore, is to set Burt's early work on factor analysis in context, in the process providing an account which revises commonly held, accepted, and incorrect views as to priorities in this field. In so doing, I expect to raise hackles and provide the opportunity for others to criticize my account. This will be all to the good. Existing accounts are in general partisan, partial, parochial, self-serving, or else aimed at providing a painless introduction to the tyro. I have no personal connection with any of the protagonists in the story I set out to tell—indeed the squabbles and backstabbing that seem so inevitable a part of academic rivalry and which are the constant subtext of this chapter are one of the least attractive aspects of academic life. My own original technical contribution to the field is limited to a single paper at a conference of the Psychometric Society which I never wrote up for publication and which, I now realize in the aftermath of researching material for this chapter, was a modern solution to a problem raised 60 years earlier by Stephenson (a collaborator of both Spearman and Burt).

A good part of my unseen contribution to current research has been as a reviewer of the work of others prior to publication, partly as a result of which I have become increasingly sceptical of the value of factor analysis in the hands of the non-specialist. Indeed, I date the decline of moral standards in Western society to the mid-1960s not, as it happens, because of the advent of the contraceptive pill, but because of the growth in availability of canned factor analysis routines in university computer centres. So the only axe I have to grind, though grind it I will, is aimed at the necks of those who seem to me to misrepresent the logic of the method, the origins of modern practice, the utility of various techniques, or the connections between technical choices and substantive theories in psychology.

It is the case that there has been considerable confusion from the very first years of this century between technical issues in factor analysis and substantive theories of the organization of psychological functioning. Seen from the perspectives of the original researchers this is not only understandable, it is perfectly excusable. What troubles me is that an uncritical version of how factor-based theories and factor analytic methods were developed has passed from textbook writers through generations of university lecturers into the examination answers of generations of university undergraduates, and so into such popular imagination as contemplates the matter, without passing through the minds of any of them. I hope to unpick some of the knitting, and in the process to resurrect approaches and issues that have something to say to present-day concerns.

Plan of this chapter

A straightforward chronological account would too easily lead to confusion. Instead, I shall first set the context of Burt's early work, paying particular attention to Spearman and the issues his theory raised, and drawing attention to J. C. Maxwell Garnett, a largely neglected figure whose contribution has been much underrated. Then we shall look at what Burt himself did, identifying the novelty of his contribution. The elimination of Burt from the written history, the new flush of work from America, and the relationship between the work of Burt and that of Thurstone is the topic of the penultimate section, followed by some conclusions.

A note on culture

In examining the origins of factor analysis, it is almost necessary to don the mantle of an archaeologist rather than that of an historian. A lot of relevant material has not survived. True enough, the protagonists in this tale wrote in English, but it was a language in transition even then, and more of the resonances of the vocabulary used have been lost over the intervening years. The statistical notation used was often cumbersome, and the language of mathematics then used has become ever more obscure. Assumptions concerning standards of presentation and bibliographic conventions, not to mention social attitudes, were different and interpose themselves between the present-day reader and his subject matter. The content of early papers can seem by turns quaint, insightful, and offensive to the modern eye. Spearman and Hart (1912) cheerfully suggest that adequate scores on intelligence tests may one day be a requirement for voting and even breeding rights, for instance, while Burt (1909) talks about the heritability of intelligence from father to son (rather than in terms of the transmission of genetic material) and without considering the mothers or indeed grandparents.

The correspondence between Burt and Spearman preserved in the Burt archive at the University of Liverpool also illustrates how cultural change makes it difficult at times to make sense of the record. I have come to the view that Burt and Spearman in later years were corresponding at cross purposes through a fog of insincere pleasantries. For instance, in 1937 Spearman writes to 'Professor C. Burt, F.R.S.', presumably knowing full well that Burt was not a Fellow of the Royal Society, although he himself was. How much of the formality of their correspondence, and in particular the obsequiousness of Burt's letters, should be attributed to the common culture of the time, and how much to the state of their relationship, I cannot tell.

In the period we shall consider—from 1901 to 1940 or thereabouts—computers were people who made a living undertaking repetitive mechanical arithmetic according to instructions provided. Short-cut formulae, reasonable approximations, and proofs that terms in equations vanished asymptotically, were valuable contributions to the efficiency and timeliness of research output. The existence of such short-cuts has probably been the single most obstructive hurdle in the path of beginners trying to get to grips with the kinds of statistics psychologists use, as the natural shape and elegance of fundamental formulae are supplanted by ungainly and far less comprehensible computationally efficient surrogates.

In particular, given the comparatively settled state of multivariate analysis techniques today, it comes as something of a shock when Burt's presentation of a correlation matrix is accompanied by a detailed description of how such a matrix should be read (Burt 1909), as if this mode of presentation were completely novel. It also comes as something of a shock on reading his mature account of factor analysis (Burt 1940) that his understanding of it is deeply imbued with the intellectual machinery of Aristotelian logic—a fact which casts light on the differences between him and Spearman.

Origins

From the point of view of most psychologists, factor analysis dates from the publication of Spearman's celebrated paper 'General intelligence: objectively determined and measured' (Spearman 1904b). From the point of view of the rest of the world, the origins of the techniques *now* used in factor analysis and related areas are to be found in 'On lines and planes of closest fit to systems of points in space' published in the *Philosophical Magazine* in 1901 by Karl Pearson (Pearson 1901). It has to be said from the outset that Pearson's paper, although it is undoubtedly the original statement of the mathematics of the principal axes method, had no direct impact on the development of psychological work in factor analysis for some considerable time. Spearman does not appear to have read it, or if he did it had no influence on his work. Perhaps this is not surprising. A long-running acrimonious dispute festered between Pearson and Spearman concerning certain criticisms Pearson made of Spearman's paper on correlation (Spearman 1904a), and extending to various other technical issues. For those interested in sampling the flavour of their relations, Spearman

(1928) is perhaps one of the most easily accessible (in the physical sense) papers.

However, Burt (1952) claims that his original interest in factor analysis was sparked by a paper in *Biometrika* (Macdonell 1901) containing a suggestion by Pearson that correlations amongst certain physical variables might be appropriately analysed in the way he had proposed, and by a talk given by Pearson at a college society in which he outlined his ideas. Burt further claims to have outlined to a group of his fellow undergraduates his scheme for deploying a simplified version of Pearson's methods on psychological measures, and to have been the butt of salutary criticism from his fellows.

Nothing other than Burt's own account of this early foray now remains, but the papers by Pearson and Macdonell are there for all to see, and contain precisely what Burt claimed they did. For a good account of Pearson's work expressed in modern terms see Krzanowski (1988, p. 53 *ff.*). Pearson's method of principal axes re-emerged in the 1930s once efficient computational algorithms had been devised, and has been central to factor analysis every since—but more of this later.

In the Burt–Spearman correspondence, Burt lays stress on his indebtedness to Spearman, and discounts Pearson. On the surface, this tends to support the view that Burt's claim to an intellectual inheritance from Pearson was concocted after Spearman's death. But on the other hand, given the bad blood between Spearman and Pearson, it is hard to imagine anyone writing to either of them emphasizing the influence of the other.

Spearman

Spearman was the first to use the term 'factor' in anything like its modern sense, and in that sense can properly be considered the father of factor analysis (or factorial analysis as he more usually termed it).* 'Factor' turns out to be a weasel word, by accident of usage rather than design. So far as Spearman was concerned, the general factor he found and named g was in some sense an active psychological principle, and he spent most of his academic career elucidating its properties, attempting to identify its basis, elaborating laws concerning its functioning, and with increasing ingenuity defending it against all-comers. It so happens that more modern approaches to factor analysis as a technique for examining the structure of correlations amongst variables can be cast in terms of matrix algebra so as to make the word 'factor' a mathematical term of art as well as an indicator of some real causal influence. You factor a correlation matrix by expressing it as the product of a matrix of factor loadings and its transpose: the trick is to reproduce the original matrix as nearly as possible from a matrix of factor loadings with as few columns ('factors') as possible.

---

* But that is not to say that the kind of analysis Spearman undertook is the lineal ancestor of modern practice.

This is a continuing cause of confusion. From a statistical or mathematical point of view there is nothing in this procedure that guarantees the uniqueness, or meaningfulness, or interpretability of the results. But it is all too easy to slip sideways from the pronouncements of a standard computer package as to how many and what factors are to be found into the supposition that the factors so determined are likely to be real psychological influences. The best cautionary text I know on this topic is Burt (1940), whose discussion of the metaphysical status of factors remains the most erudite and subtle I have read. His words of caution could be read with profit by many present-day investigators.

But Spearman knew nothing of this future usage. He did not treat sets of correlations as matrices, nor did he employ matrix algebra in his accounts of his work. The original identification of g was made in a different fashion (although it would these days be possible to construct an analogous demonstration in matrix terms).

The crucial step that Spearman made was to recognize that the correlation between two measures is depressed by errors of measurement. Making what he considered to be the appropriate adjustment using scores on a handful of measures with schoolchildren, he was, one surmises, startled to find that correlations between apparently disparate measures turned out to be more or less perfect. That he was startled comes across in the tone of his account of this work (Spearman 1904b): he clearly considered he had made a discovery of the first importance. Tyros are frequently startled by factor analysis (and everyone was a tyro in 1904): lurking in the complexities of a large matrix of correlations may be regularities one little suspects, simply for lack of practice at inspecting such matrices.

It is worth pondering the novelty of what Spearman proposed. In so far as apparently disparate measures of intellectual functioning are correlated, this can be accounted for by one, and only one, and always the same general factor, viz. general intelligence. Up to this time, the search for structure in the correlations amongst what were taken to be appropriate measures had been disappointing (e.g. Wissler 1901). It is at least possible that Spearman's apparent success where others had failed can be accounted for by the fact that his subjects were not university undergraduates. Restriction of range is another potent depressant of correlations: those whose work on intelligence is limited to intellectually able groups are less likely to develop a respect for the omnipresence of g than those who strive to produce g-free or g-reduced tests for use across a wide spread of ability.*

According to Gould (1981, p. 315):

The fact of pervasive positive correlation between mental tests must be among the most unsurprising major discoveries in the history of science.

* This last sentence was written with some feeling: the production of a battery of tests for occupational selection purposes which correlated little amongst themselves but well with later work performance would be the route to a gold-plated retirement.

The truth is that in 1904 it was surprising, or unexpected, or at the very least it had not been discovered before, despite strenuous efforts to establish some pattern among plausible measures of intellectual functioning. Spearman's career was built on the discovery. Great interest and controversy was aroused by his results. It is only unsurprising now because the intervening century has yielded myriad instances of the same result.

It is also worth reading Spearman's original paper—not in the truncated form in which it sometimes appears, bereft of the sums, but in the full version. Careful reading, for instance, will show that the correlation coefficient Spearman used was not Pearson's product–moment coefficient, but his own rank–order correlation, which at that time he called the 'foot-rule' correlation. This is one of the short-cut methods referred to earlier: it greatly reduces the amount of computation required, and is entirely the same as calculating Pearson's coefficient on the ranks of the data rather than directly on the data themselves.* Readers may be surprised to discover how few subjects he used and how large the claims he was prepared to make on the basis of modest amounts of data.

None the less, this paper set the agenda for 25 years and more of academic controversy, not just because of the claims it made for the discovery of a uniquely important principle of psychological functioning—others were doing that in books and papers which contained not a single statistic—but because it proposed a criterion which was amenable to computation and therefore to public debate. This is the notion of a hierarchy among correlations.

The hierarchy

According to Spearman's interpretation of his results, measures of intellectual functioning differ in their 'saturation' with g. Those measures which are more saturated with g correlate more highly amongst themselves, precisely because it is only on account of their saturation with g that they correlate at all. So, if g is all that is needed to account for the correlations amongst a set of measures, it should be possible to arrange the measures in order, from the most g-saturated to the least, and observe a steady decline from most to least in the size of correlations.

This Spearman called a hierarchy. It is not, I think, the most obvious term, and in view of later developments it invites confusion. It has nothing to do with the notion developed by Burt and by Vernon of factors at different levels of generality which can be depicted as a hierarchy in the form of a tree-diagram. Nor does it relate to the idea of first- and second-order factors in Thurstone's style of multiple factor analysis. It is simply that one can arrange a set of correlations in a particular pattern, which can be recognized by eye.

* Burt also used a version of Spearman's coefficient in his first paper (Burt 1909).

General factors and the positive manifold

A word is in order at this point about the omnipresence of general factors and the overwhelming preponderance of positive correlations where tests of human ability are concerned. It is not true to say that factor solutions always yield a general factor. With a little care in the choice of variables, one can contrive matrices where some small proportion of the variables have negligible loadings on the first factor, and I have seen a number of such matrices in the course of my career. Nor is it the case that general factors always load positively on all variables—again, a change in scoring conventions can easily be arranged, and with measures other than ability tests (for instance scores from personality questionnaires) bipolar first factors are by no means unusual.

But so far as measures of human intellectual ability are concerned, assuming that higher scores are associated with better performance, the evidence is overwhelming that the correlations are positive. For this reason, a general factor in an analysis of ability tests is liable to be identified with general intelligence by those disposed to be interested in such a concept. A corresponding tendency has been to suppose that the second factor, being inevitably bipolar under such circumstances, has some special status by reason of the presence of minus signs. But because of the controversies which have raged around ideas concerning intelligence, IQ, and intelligence testing, it is too easy to overlook the delicate distinctions which writers on factor analysis have made from the earliest times, and to suppose that they all leaped uncritically to the conclusion that the general factor extracted from a battery of tests should be immediately identified with intelligence, although of course Spearman's first shot was to do just this.

The substantive finding that ability tests correlate positively and the mathematics of early methods of factor analysis intertwined in such a way as to make such distinctions seem nugatory to many critics. Just as many American writers assumed quite wrongly that a single British school of factor analysis existed, led by Spearman, and ignored the distinctive contributions and methods of Thomson, Garnett, and Burt, there has been a corresponding simplifying tendency to identify general-factor techniques in factor extraction with particular theories concerning mental organization.

Thomson and Garnett

The more precise definition of a hierarchy, and the question of whether the existence of a hierarchy necessitated the presence of a general factor (and vice versa) eventually became a matter of heated dispute. Godfrey Thomson (Thomson 1916) constructed data with a hierarchy but no general factor and went on to elaborate his competing account of mental organization over many years (e.g. Thomson 1935). With Burt he co-founded and co-edited the *British Journal of Statistical Psychology*. He appears to be little read these days, which is a pity. J. C. Maxwell Garnett (1919*a*, 1919*b*, 1920, 1921) provided proofs relating the existence of hierarchies to the presence of general factors and vice versa, introduced the idea that correlations may be treated as the cosine of the angle between two vectors—a crucial idea in the development of a geometrical model of factor analysis—laid stress on the existence of factors other than the general factor (so-called group factors, of which more later), and thoroughly annoyed Spearman by repeatedly referring to the proportionality criterion (of which more very shortly) as 'Mr Burt's equation'. He then disappeared from the literature totally, resigning his post as Principal of the Manchester College of Technology to become secretary of the League of Nations Union and grandfather of Peter Jay, economist and sometime British Ambassador to Washington, and Virginia Bottomley, at the time of writing Secretary of State for Health.

Garnett had been a lecturer in Pearson's department at one time: Burt (1947) credited him with the first clear statement of the fundamental method of multiple factor analysis as later (and now) practised. Modern texts in general do not refer to Garnett, although Hearnshaw (1979) mentions him briefly, and Boring (1950) recognizes his importance. An exception is Guilford and Hoepfner (1971), where (p. 2) he is credited with being 'perhaps the first to break with the prevailing doctrine' and emphasize group factors, whereas Thurstone is credited with 'popularising' multiple factor analysis.

In fact, Garnett's work has been unaccountably ignored for about 60 years. Originally a mathematician, he did work in the solid-state physics of ruby which is still read today because, I am given to understand, it relates to the development of lasers. He was called to the Bar, then joined the Board of Education, and was appointed to be Principal of the Manchester College of Technology (now the University of Manchester Institute of Science and Technology) in his early thirties. His departure from that post was occasioned by a dispute concerning the admission of degree-level students, of which he was strongly in favour. This dispute appears to have become very heated indeed.

Garnett's career as a factor analyst lasted no more than a handful of years, and his influence on psychologists was perhaps limited by the fact that he was a professional mathematician. His most important paper (Garnett 1919*a*) is very heavy going, but it is, as Burt later pointed out, the first enunciation of certain important proofs and marks the inauguration of multiple factor analysis formally described. For instance, he shows that the correlations among just three variables cannot establish the existence of a general factor, and that the hierarchy, which was crucial to Spearman's approach, is a special limiting case of a more general pattern involving overlapping group factors. He reanalyses Webb's (1915) data to show the presence of factors other than g, and lays the foundations for the spatial–geometrical interpretation of factors which dominates introductory textbooks to this day. Garnett's (1921) book *Education and world citizenship* has the distinction of containing not only a plan for the future organization of education in a world context, but also, in Appendix B, a technical account of factor analysis as then understood.

Garnett drew heavily on Burt's work, and was impressed in particular by his empirical findings. They were close contemporaries, and may well have met in the context of the so-called 'social settlement' (a sort of experimental community) in south London, with which they were both involved. Given Burt's rôle as psychologist to the London County Council and Garnett's job, which related to the education of young working-class people, it would be surprising if they had not corresponded: the discovery of any surviving correspondence has become an (as yet unsatisfied) ambition of mine.

Spearman's interest, first and foremost, was in the general factor he had discovered, in elucidating its nature and elaborating its psychological importance, whereas Burt, almost as soon as his first paper was published, was pursuing the application of factor analysis as a technique. As practised in Spearman's group, factor analysis involved 'purging' or 'purifying' batteries of tests, dropping those which disturbed the pattern of the hierarchy, so as to get an ever more accurate fix on the nature of g. Burt went on to develop his approach to factor analysis applying it to different kinds of measures, including measures of emotional variables, and evolving a distinctive and original style of analysis.

The literature on factor analysis between 1910 and 1920, leaving aside Burt's contributions, is largely directed to questions concerning the reality or otherwise of g and the relationship between hierarchy among correlations and the presence of a general factor. The hot topic was general intelligence, as revealed by factor analysis, not factor analysis and how it could be put to work solving many and various problems. None the less, there is a good deal of tedious algebra to read. These matters occupied those British psychologists interested in the factorial account of intellectual functioning for many years. Anyone who thinks acrimonious exchanges in scholarly journals are a recent phenomenon need look no further to be disabused.

Where was Burt in all this? He was listed as an assistant editor of the *British Journal of Psychology* soon after the appearance of his own first paper, and may be presumed to have been aware of the various controversies that decorated its pages. His employment history is also well documented. On my reading of the evidence, he was developing his own line of thinking and keeping out of the main scuffle, whose protagonists chose to ignore his techniques (except to attribute an equation to him). He may have been overawed by their mathematical sophistication.

Burt's first paper

I believe it to be the case that Burt still holds the record for the longest paper ever to have been published by the *British Journal of Psychology*, and this at his first attempt (Burt 1909). It was, on his account (Burt 1952), to have been even longer, but an appendix of mathematical material was not printed because both Pearson and Spearman raised objections. It is Burt's claim that this appendix contained a first account of his method of 'simple summation', the development of which is one of the main matters in contention concerning his claims to a significant place in the history of factor analysis.

The paper exists now only in its published form. Correspondence between Burt and Spearman has survived from 1909, but unfortunately the draft on which Spearman made extensive comments has not. In one letter, Spearman asks for a note to be added (though which I cannot tell) because 'It is the sort of thing a man like Pearson might try to get hold of,' which perhaps gives the flavour of the Pearson–Spearman rivalry. In 1937, Burt writes to Spearman:

Karl Pearson, who read my draft and strongly criticized all the mathematical methods, observed . . .

and

This latter and 'other mathematical technicalities' were subsequently dropped (as a footnote explains) since McDougall very wisely said . . .

The content of this letter is consistent both with the contents of Burt's early work and with his later claims concerning simple summation. It predates Hearnshaw's notion of when the 'new story' first appeared by about ten years.

This 1909 paper was the first in the history of psychology to attempt to test a factorial hypothesis by fitting an explicit model to new data. The model was Spearman's, the method, so far as one can tell, was Burt's. What he did was this. He collected data on a variety of intellectual measures from boys in various schools (the nature of the measures, and the identity of the schools, are of interest from the point of view of the more general history of the study of intelligence, but do not touch our present concerns). Then, as we noted above, using Spearman's rank–order correlation coefficient, he obtained matrices of correlations the manner of reading of which he explained at some length. He then worked out what the theoretical correlations ought to have been under Spearman's hypothesis of a single general factor, and produced a table showing observed and expected correlations, and the probable error and the deviation (that is, the difference between observed and expected correlations). The tables as printed, especially Table V, contain a number of errors; these appear to be typographical errors rather than errors in computation, since comparison with the corresponding elements in a summary table at the end of the paper allows corrections to be made, and the nature of the errors suggests that the compositor was having a bad day.

He did not calculate an overall goodness-of-fit test between observed and expected correlation matrices, but that is scarcely surprising since no such test had yet been developed (although—again on his own account (Burt 1952)—he claims he had been prompted in this direction by Pearson's work on goodness-of-fit in contingency tables). He noted that there was a strong overall tendency for the correlations to fall into a hierarchy in the manner Spearman had proposed, but he also noted, though rather mutedly, that some appreciable deviations existed between observed and expected correlations. Joynson (1989, pp. 92–3) reprints Table V, complete with errors.

From the surviving correspondence it is clear that Spearman read, commented upon, and influenced this paper before publication. Burt (1952) would have us believe that Pearson did as well. He also claims that whilst both were critical of his work, they were even more critical of each other, which is not difficult to believe. If it is true, as Burt claimed, that he produced what each of these considered dubious mathematical procedures, to the extent that they were never printed, and that these procedures were essentially those that later came to be called 'simple summation', one can understand his reviewers' reluctance. On the one hand, simple summation is the kind of procedure that gives real mathematicians indigestion: it delivers at best an approximation to the major axis of the correlational ellipsoid that Pearson had discussed in 1901, and this approximation is easily perturbed by the addition or deletion of variables. On the other hand, emphasis on the differences between the theoretical and observed coefficients tended to undermine Spearman's position as regards the sole importance of his general factor. The relevant documents detailing the criticisms have not survived.

When Burt sent a draft of his paper, including a hierarchy of theoretical correlations, to Spearman for comment, before submitting it for publication, Spearman supplied various formulae for calculating theoretical values and for judging whether a hierarchy was to be found. To this latter end, Spearman proposed what has become known as the proportionality criterion, which with a little simple algebra can be transformed into the 'tetrad difference equation' which became a principal tool in Spearman's style of factor analysis. This equation has led a hectic life, and been the focus of some considerable controversy in the matter of priorities.

Mr Burt's equation

The proportionality criterion may be stated thus:

$$\frac{r(A,P)}{r(B,P)} = \frac{r(A,Q)}{r(B,Q)} .$$

That is to say, the ratio of the correlations between any two variables and a third variable is a constant. Applying this criterion to a complete set of intercorrelations gives an indication as to whether they form a hierarchy in Spearman's sense. There is no doubt that this formula was supplied to Burt by Spearman, that Burt's use of it in his 1909 paper was the first formal publication, and that he pointed out in a footnote that it is immediately deducible from an equation in Krüger and Spearman (1906). It became known as 'Burt's equation' at least partly because Spearman and Hart (1912) called it Burt's equation. Garnett (1919b) made repeated and intensive use of the expression 'Mr Burt's conditions'—so frequent and intensive that I found it curiously jarring. By this time Spearman had clearly developed proprietorial concern for it and raised objections, and Garnett published a correction pointing out the true origins.

None the less, in his letter of 23 June 1909, Spearman writes to Burt, 'using your notation', and then gives the formula as above. It may be that Burt's notation was used by Spearman because it made the structure of the equation particularly clear (permitting a simple one-sentence statement as above, which is not true of the equation from which it was derived).

Why, one wonders, was Spearman once so careless as to priority in this matter, and later so jealous? The matter re-emerged nearly 20 years later when Burt (1937) stated that the criterion was first 'given' in his 1909 paper. Spearman cavilled at this and Burt accepted his priority. Much has been made of this incident—more, I think, than it has deserved. Burt never seems to have claimed that he invented the proportionality criterion, only that he was the first to publish it formally, which is manifestly true. It may well be that he chose his words very carefully indeed to suit his purposes, but he would not be the first academic to do that. In 1937 Burt writes to Spearman:

When you substituted 'equation (f)' for 'equation (a)', I took it that I had made a slip of the pen. I now see that my own version of your criterion could not have been deduced from (f) . . .

This raises the entertaining possibility that all of Burt's early work in factor analysis, leading to the simple summation method, the delineation of special abilities in educational and industrial psychology, and so forth, was based on a misunderstanding of Spearman's work.

It is not just that Burt published the equation, he used it to test the fit of his observed correlations to what theory would predict. I have puzzled over his reticence in pointing out the lack of fit, Spearman's and Pearson's reported reservations about the unpublished mathematics, and Spearman's use of the term 'Burt's equation' for a formula for which he later wanted exclusive credit. My guess is—and it is only a guess—that Spearman did not at first realize the use to which Burt would put the equation. It is a subtle difference, but Spearman's emphasis was on showing how well the hypothesis of a single general factor fitted the data; Burt's calculation of deviations from expected values too readily laid stress on badness of fit. For a time at least thereafter, Spearman emphasized other criteria for determining the existence of a hierarchy, and in particular the progressive diminution in size of correlation coefficients from the upper left to the lower right corners of an ordered matrix. In fact Spearman himself did not adopt Burt's approach to assessing the adequacy of fit: for many years he used the correlations between the columns of a correlation matrix for this purpose, and then in the mid-1920s he turned to the tetrad difference equation. Perhaps it suited Spearman to distance himself from the strictest criterion for the presence of a hierarchy when data suggested the hierarchy was less than clear. Perhaps Burt was trying to pull a fast one, and was reined in by Spearman and Pearson in their different ways. Both Burt and Spearman, from their respective points of view, had good reason to want to see the paper published in some form. As printed it may plausibly be regarded as an uneasy compromise between what Burt wanted to write and what Spearman wanted to read.

Analysis of residuals

Burt went on to develop his own brand of factor analysis. It is generally agreed that by 1917 he was, in a variant of his method for calculating 'deviations' in the 1909 paper, at the point of identifying factors other than the general factor in correlation matrices (Burt 1917). Here I choose my words carefully. He was not extracting several factors, as he would later, but he was extracting a general factor and paying close attention to the residuals. He used the same procedure (Burt 1915) in an analysis of measures of emotion. Bickersteth and Burt (1916) present a preliminary version of material later to appear in Burt (1917), and Carey (1916) makes clear reference to the identification of factors other than the general factor in this paper, as a result, it would appear, of advice from Burt in his editorial rôle:

With actual tests of school abilities, Mr Burt has found a 'general educational ability', apparently combined with specific educational abilities (arithmetical, linguistic, literary and manual) similar to the combination of the 'hypothetical general factor' with subordinate group-factors in his intelligence tests.

In a departure from his 1909 method, he presented the deviations not as simple differences between theoretical and observed correlations, but as partial correlations, that is to say as correlations between tests adjusting for the effects of a general factor. I surmise that there was a good reason for this small change of tack. The residuals rescaled as partial correlations look bigger than they do when expressed as simple residuals. For instance, the difference between observed and theoretical values for the correlation between reading speed and reading comprehension in his data is 0.15, but the partial correlation, or to use his term the specific correlation, is 0.36. This presentational ploy dramatizes the existence of effects that cannot be attributed to the general factor. None the less, Burt's original method is the natural way of proceeding in multiple factor analysis.

In both the 1915 analysis of measures of emotion and the 1917 analysis of correlations among school subjects, Burt first extracts a general factor, and then points to a surprising pattern in the partial correlations. Just as the original correlations can be ordered in such a way as to demonstrate a hierarchy, the partial correlations can be ordered to demonstrate a cyclic pattern. This is a discovery of Burt's whose priority has never been properly recognized. Most usually, the description of simplex, circumplex, and more general radex patterns amongst correlations, where correlations adopt values corresponding to their relative position in a matrix, is credited to Guttman (1954).

In a simplex pattern, correlations decrease progressively the further they are from the diagonal. This pattern is common in correlations amongst items in ability tests when the items are arranged in order of difficulty. In a circumplex pattern, correlations first decrease, then increase once more the further they are from the diagonal, so that a circular pattern of relationships is suggested. More complex patterns are subsumed under the name 'radex'. Spearman's hierarchy is of course another example of a matrix following a simple pattern, but it is quite plain from Burt (1915) that he recognizes the circumplex nearly 40 years before Guttman.

That his thinking is developing along lines quite independent of Spearman is illustrated by the following extract from Burt (1915):

This phenomenon—the cyclic overlap of specific factors—has also been observed in partial coefficients obtained by eliminating 'general intelligence' from the intercorrelations of complex mental tests. It may, therefore, have a wide significance for the study of mental organisation.

The theory of a general factor is commonly associated with the views that specific factors are negligible and that the general factor is simple. The problem arises as to whether the above specific correlations invalidate the hypothesis of a general factor, and, in particular, whether the peculiar relations between the specific factors may not of themselves produce the appearance of a general factor.

In fact, this style of analysis of the residuals after extracting a general factor has been largely ignored in the subsequent development of factorial approaches to individual differences. It is easy to see why. Burt's own direct concerns led him to have a greater interest in the possibility of defining 'group' factors. Once Thurstone's style of multiple factor analysis*

---

* By multiple analysis is meant the extraction of more than one non-specific factor from a set of correlations. Spearman had a 'two factor' theory in which the general factor combined with a specific factor in determining a particular kind of performance. In this theory there were no 'group' factors shared by a subset of disparate measures. Multiple analysis allows such factors.

gained favour, there was no motive among his followers for supposing that the first factor extracted had any special status, so no reason to suppose that the residuals might follow any particular pattern. And although Guttman (1966) showed neatly that the conventional form of multiple factor analysis grossly overestimated the number of dimensions needed to account for simplex and circumplex patterns among correlations, interest in this style of analysis has been desultory. More recently, with the widespread availability of prepackaged factor analysis routines for digital (as opposed to human) computers, the notion that one should take any account at all of the residuals seems to have disappeared from view.

But the step from recognizing that factors beyond a general factor need to be taken into account to achieving a full-scale multiple-factor analysis is neither obvious nor straightforward.

In retrospect Burt (in Burt 1940) and other writers were able to recognize the interrelationships amongst various alternative approaches, but at the time progress was hampered by a theoretical confusion and a computational hurdle. The theoretical confusion—the unspoken assumption amongst those doing various forms of multiple analysis, e.g. Garnett, Kelley, and Holzinger, that factors defined on the correlation matrix were also factors as organizing principles of mental life—led to much emphasis on unrotated factors as extracted by whatever means.

The computational hurdle was more teasing. General factor methods such as Burt's method of simple summation involved calculating the square root of the sum of the correlations in the matrix. Indeed, what may be called the unique selling point of simple summation is that this is the most difficult arithmetical operation that has to be performed. Simple summation proceeds by adding up the elements in each row (or column) and dividing by the square root of the grand total. Once the first factor has been extracted, the sum of the residual correlations is precisely zero, some being positive in sign, others negative (this is not a matter of empirical observation but of mathematical necessity).

Working with the square root of zero does not take one far in mathematics. Opinion differed as to the best way to proceed, and as late as 1940 Burt was still offering advice as to how best to decide which method to use. For instance, one may be able to identify subgroups of variables which have appreciable residual correlations with each other, but negligible correlations with the rest; or one can systematically change the signs of the residual correlations, do the calculations, and then restore the original signs.

Burt in particular, of those who remained active in the field after the first flush of debate, was insistent on the importance of group factors. On the one hand, he saw that empirical data demanded some account of non-negligible residuals after the extraction of a first factor. On the other hand, he developed the view that special abilities, to use one of his terms for

group factors, could be trained, and therefore were of particular importance to educationalists.

Another note on culture

Burt's use of the term special abilities raises a point of terminology that is of
no little importance in the early history of factor analysis. Spearman
claimed that performance on any task involving intellectual ability was
determined by the general factor g and a factor specific to each task s. It
may be just an accident of history that he chose those terms, but the choice
of a proper means of describing his findings led Spearman into 30 years of
dispute as to what he meant by specific. In general terms, he was surely
trying to exclude the existence of precisely those abilities termed 'special',
or group factors, whose properties Burt sought to elucidate.

In his textbook on factor analysis (Burt 1940), Burt made the claim that
factor analysis was no more than an extension of logic. There are a number
of ways of interpreting this claim. It could be a simple means of soothing
the nerves of the daunted student facing an impenetrable wall of jargon; it
could be an echo of Burt's interest in Russell and Whitehead's claim to
have shown that mathematics is merely a continuation of formal logic. But
I think it is neither of these. Burt read Greats at Oxford, and in his time this
involved exposure to the traditional logic, derived ultimately from Aristotle
by way of Porphyry and elaborated by the medieval scholastics. In this
style of logic, which was part of the patrimony of Western culture for
several hundred years up to perhaps the middle or later years of this
century, an important rôle was played by the fourfold scheme of predicables, *genus, species, proprium*, and *accidens*. To anyone schooled both in
Latin and this system of logic (and by an accident of education, though by a
different route, I share Burt's background in these respects), words like
general, specific, special, proper, property, accident, and so forth have not
merely a ring of ancient etymology, they can take on the aspect of technical
jargon. It is all too easy to slip between the modern everyday use of such
words and the original technical usage. For an example of such slippery
usage see the previous paragraph.

In Burt (1940) the connection between the predicables and different kinds
of factors is made very explicit. For *accidens* read error variance; for *proprium*
read reliable variance associated with a single test; for *species* read group
factors; for *genus* read general factor. Or at least that gets somewhere near
the scheme Burt wants to adopt. According to this way of thinking, the
typical bipolar second factor represents a *distinction*, not a causal agency,
and the process of factor analysis is the logical decomposition of a correlation matrix aided by arithmetic rather than a fixed mathematical
procedure.

At this distance in time it is difficult to determine to what extent he made consistent technical rather than vernacular use of these words in his early years, but there does seem to have been a widespread lack of agreement in particular as to what words like 'special' and 'specific' meant in the context of factor analysis. In the end, Burt's 'special abilities' became 'group factors', and Spearman repeatedly glossed his own usage. But the older technical meanings seem to have driven some of Burt's best ideas about the conduct and interpretation of factor analysis. The content of his early writings is consistent with his later claim that he was using the words 'special' and 'specific' in the technical sense, but in the confused climate of an inchoate discipline it is always possible that he was less precise and definite than he would later have wanted us to believe.

The dog that did not bark

Until Hearnshaw's biography appeared, there was no suggestion anywhere that Burt did not devise the method of simple summation as an approach to factor analysis. Burt details the method in *The factors of the mind* (1940); it is there to be found in Burt (1917), although one needs a practised eye to see it. From the perspective of 80 years ago, it might have seemed like a rough approximation to 'proper' factor analysis, which used formulae involving many more computational steps. From the perspective of the 1940s it turned out to be precisely what Thurstone termed 'the centroid method'. Between the publication of *The factors of the mind* and Hearnshaw's biography there was ample opportunity for Burt's claim to have originated simple summation to have been challenged. It was not. It does not take the deductive power of Sherlock Holmes to come to the conclusion that no one complained (not even Spearman) because no one thought his claim was false. Stephenson (1984) has this to say:

. . . Spearman could indeed become very angry, as he did when I brought group factors *v* and *k* to his notice; he considered those to be intrinsically anoetic, and thus not what 'nature' prescribed as noesis. Thus he was in no mood to consider multiple-factor or summation methods of factor analysis: moreover I am of the opinion that this was fully justified.

Although it is clear, therefore, that Burt was doing factor analysis by simple summation well before 1920, the trouble is that he did not stop to set out a detailed description of the procedure. But later editions of *Mental and scholastic tests*, first published in 1921, contain as Appendix V to Memorandum II just such an account, identified in a footnote as having been originally a report to the Education Officer of the London County Council in 1920. And here begins a tale as confusing as any in the whole Burt saga.

In the course of my research for this chapter I consulted three different copies of *Mental and scholastic tests*. London University Library holds copies of the 1947 and 1962 printings (I use the word advisedly) on open shelves, and for convenience I used these mostly in my research. The Greater London Record Office has a copy of the original (1921) edition, which has not seen much use, and which I consulted early and late. In point of fact, the book was first published by P. S. King for the London County Council, as was Burt (1917). The London County Council minutes record the decision to publish, with some enthusiasm for the importance of their Psychologist's work. But the 1921 edition does not contain Appendix V to Memorandum II. Steinberg's (1951) bibliography of Burt's publications, listed in chronological order, has as item 44:

Mental and Scholastic Tests, Second ed. (PS King and Son, 1927, pp. xvi+447). A revised reprint of the first edition with further appendices on factorial analysis and its results.

But the 1927 printing (consulted by N. J. Mackintosh), dated November 1927, is described as the third impression of the first edition, and does not contain Appendix V. The 1947 edition declares itself to be the first printing of the second edition and does contain Appendix V (it also lists fourth and fifth impressions of the first edition as having been printed in 1933 and 1939). But the 1962 edition proclaims itself to be the fourth edition, with a second edition in 1933 and a third in 1947.*

So Appendix V did not in fact appear until 1947, although it is supposed to have been written in 1920. I have been unable to find any record of the 1920 report in the London County Council archives, or in the London County Council collection in the British Library. The London County Council minutes make no reference to it, although the Education Committee notes many matters in meticulous detail, including the proposed reduction in the Psychologist's salary from £450 to £350 per annum as a post-war retrenchment measure in 1920. There are no London County Council papers in the Burt archive for the period in question.**

* *Note added by the editor*
The bibliography published by Steinberg in 1951 was described as prepared on behalf of the University College London Psychological Society on the occasion of Burt's retirement, and mentions in a footnote that 'explanatory notes have been added' for some of the earlier and less accessible publications. According to Professor Steinberg:

> The list of publications came about because Burt was retiring and we students thought that an annotated list would be a suitable tribute. It fell to me as president of the students' psychological society to ask Burt for a full set of his reprints from which we intended to make annotations. Burt was pleased and agreed but I was surprised when he handed me a complete typescript with my name on it and no reprints, and said cheerfully that it had all been done. C. W. Valentine, an old friend of Burt's, wanted to publish the list in the *British Journal of Educational Psychology* of which he was editor. I saw the manuscript through to publication.
>
> (Personal communication to N. J. Mackintosh, October 1994)

** A footnote in the 1962 edition of *Mental and scholastic tests* suggests that the mathematical core of this appendix was itself taken from Appendix 1 of the Annual Report of the Psychologist to the London County Council for 1914. A copy of this report is extant in the Burt archive. There is no Appendix 1.

It is difficult to imagine that the County Council would object to the inclusion of a technical appendix. So the question arises as to why, or perhaps whether, it took 26 years to find its way into *Mental and scholastic tests*, the tentative answer to which we must delay for a while. Let me make it plain that even though it is not possible to detail in Burt's publications a documented series of individual steps, beginning with common practice and ending with a clear and unambiguous technical manual for factor analysis by simple summation, it seems to me that there is no room for doubt that Burt devised it, along with discovering the first instances of a circumplex in residual data, inaugurating the practice of confirmatory factor analysis, and providing the first instances of factor extraction leaving residuals to be tested for significance. He did all of this before his 40th birthday, a decade before taking up his post at University College. It is difficult to understand why critics such as Hearnshaw, Gould, and Fancher have thought otherwise, unless they were expecting the originators of factor analysis to use post-war terminology.

What is more, according to Stephenson (1983), while Spearman's group were laboriously grappling with the tetrad equation, the cumbersome approach to significance testing derived from the proportionality criterion, Burt was elsewhere doing factor analysis 'as we know it now'. It may not quite have been 'as we know it now', but none the less, it resembles much more closely than anyone else's practice at the time both modern methods of factor analysis and the proposals of Pearson (1901).

The psychologist vanishes

In 1924 the Board of Education published its *Report of the consultative committee on psychological tests of educable capacity*. This report was heavily influenced by Burt, his findings, and his methods. By this time, in my view, such of Burt's original technical contributions to factor analytic practice as had lasting value were essentially complete. In the same year L. L. Thurstone published his book *The nature of intelligence* (Thurstone 1924). This account contains no mention of Spearman, Burt, factor analysis, or any of the research literature on our present topic. Thurstone later promoted the use of 'centroid' factor analysis, which is essentially simple summation by another name. Some writers, for example Hearnshaw, have later said that Burt 'anticipated' Thurstone, though what form of retrograde causality they are prepared to invoke I have never been sure. We now need to trace the curious story of how Burt's achievements were effectively expunged from history, a process which I have come to believe annoyed him greatly and justifiably—and perhaps led to some overstatement in later years.

Two historical accounts appeared of developments in factor analysis. Dodd (1928) in a pair of articles entitled 'The theory of factors' reviewed developments in factor analysis to date without making any mention of

Burt at all. He was National Research Council fellow in 1926–7, presumably attached to or spending considerable time at University College. He acknowledges help from Spearman, Thomson, Pearson, and Garnett. Wolfle (1940) made reference only to Burt's publications in the 1930s, and did not correct Dodd's omissions.

Thurstone (1931, 1935, 1947), once he got round to doing factor analysis, adopted a style of his own. He presented his work as if it were done entirely from first principles, eschewing a bibliography and adding only occasional footnotes by way of bibliographic material. I cannot say with certainty that he makes no mention of Burt (other than a brief and rather inaccurate paragraph in the introduction to Thurstone (1947) laying out what he believed to be Burt's mature position), simply because the footnotes tend to be abbreviated to such an extent that they are difficult to scan. Spearman and g theory are the target for Thurstone's approach to factor analysis in so far as he is critical of earlier work, and no possibility of earlier developments in multiple analysis is allowed. For instance, Thurstone (1947, p. 6) writes:

When multiple-factor analysis was introduced in 1931 . . .

Thurstone was not alone in his parochial view of affairs. Freeman (1939, p. 81) wrote:

For many years Spearman's method, and his theory of abilities which emerged from it, was the only method of factor analysis and the only theory of abilities attached to it in the field. Not all writers on the subject agreed with his interpretation but no others furnished an alternative analysis.

This view of the history of factor analysis and of priorities amongst colleagues—separated by an ocean though they may have been—is isolationist to a breathtaking degree. It is the view that is current among those who are factor analysts first and perhaps psychologists second, despite the fact that both Harman (1960) and Cronbach (1979) have attempted some degree of restitution. At first, I assumed that much of the British material was simply unavailable across the Atlantic, but the chance discovery of a reference to an article by Spearman that I had not previously read has led me to a different, and less happy view.

Spearman (1930) gives an account of factorial theory making no mention of Burt, but citing a paper by Thurstone (1925). In this paper, Thurstone reworks some of Burt's data on the revised Binet scales from *Mental and scholastic tests*, as published in 1921. By 1925, clearly, Thurstone had access to a copy of Burt's book, though without the appendix detailing the method of simple summation. In 1928 Thurstone published a paper on the estimation of absolute zero in intelligence (Thurstone 1928) in the same issue of the same journal in which Dodd's account of factor theory appeared, including a detailed account of Garnett's work. In the next year Walker (1929) published *Studies in the history of statistical method*, essentially a version of her thesis, which contains an annotated bibliography of work on factor analysis to date. Now according to Thurstone's autobiographical account (Thurstone 1952, p. 312):

The work on multiple factor analysis was started in 1929, but it did not get under way seriously for another year until completion of other commitments. The original observation equation for multiple factor analysis was written in Pittsburgh before 1922, but it was ten years before I started serious work on the problem.

Is it possible that Thurstone took no interest in Dodd's account of Garnett's work, had no access to Walker's bibliography? Was he unaware of Burt's earlier writings, did he not recognize that Burt's summation method and his centroid method are essentially one and the same, or that Garnett had already formulated multiple analysis? I find the coincidence of dates too close to call: it seems to me possible that Thurstone attended an international conference in Oxford in the early 1920s at which Burt spoke and that Thurstone's imagination was sparked. It seems to me equally possible that Burt, stung by the renown Thurstone acquired from his publications in the 1930s and 1940s, carefully manufactured a source which represented a truth: Burt was 20 years ahead of Thurstone with a fundamental formula, but had been inhibited from publishing a full account of simple summation by the criticisms of mathematically more able workers such as Pearson, Garnett, and Spearman. One can imagine how galling it must have been to see Thurstone's centroid method take centre stage when in Burt's hands it had been ignored in academic circles for 20 years, on the grounds, one imagines, that it was merely a rapid approximation with practical utility rather than serious theoretical credentials.

Whether Appendix V to Memorandum II in *Mental and scholastic tests* was really written in 1920 or many years later may never be settled, although I think there is internal evidence that may, once it has been properly analysed, swing the balance of judgement. Burt and Thurstone both had the opportunity of choosing their dates carefully in the later part of their careers, and neither's account is susceptible to conclusive proof.

The Yanks are coming . . .

In various places Burt remarked, somewhat acidly one is tempted to suppose, that 'the Americans came late to factor analysis'; one gets the feeling that he would like to imply 'and all they brought to the party was bubblegum'. In the case of Thurstone, he did bring some much-needed mathematical reformulation, introduced some prejudices concerning what might constitute a satisfactory outcome to a factor analysis, and set back the psychology of individual differences by around half a century. Whether it will ever fully recover is open to doubt. To quote Nunally (1978):

The American school of factorists has, at least until recent years, been typified by a rough-and-ready bedrock empiricism. There has not been nearly enough theory, but there has been a great deal of technical elegance in the measurement of human abilities and in the mathematical analysis of correlations among them.

For Thurstone quite explicitly wanted to discover mental faculties, and crafted his approach to factor analysis with that in view. Spearman, in introducing the notion of factors, showed how one could dispose of the need to posit mental faculties to explain differences in ability, freeing psychology from one of the drearier hangovers from its philosophical roots. Thurstone went straight back to faculties, seeing factor analysis as a method for discovering and identifying them. He asked the question, how many and what kinds of (mathematical) factors are needed to account for observed correlations among tests of ability? This was not a new question: it had been asked by Brown (1910) and Garnett (1919a) as well as Burt, and of course the running debate between Spearman and Thomson continually picked at the notion that a unitary general intelligence was implied by a single general factor.

But Thurstone's answer was to introduce criteria for the acceptability of a factor solution that remain today enshrined in shrink-wrapped computer packages: two which explicitly ruled out Burt's preferred style of analysis were the rule that every factor must have at least one zero loading, and the rule that if all observed correlations are positive no negative factor loadings are allowed, that is to say no hypothetical factor should be supposed to have a negative influence on performance. The first of these banned the general factor by *fiat*, the second ruled out Burt's analysis in terms of genus and difference in the style of traditional logic, where factors were treated as logical distinctions rather than real causal agencies.

Thurstone's prejudices have passed into current practice largely unchallenged. The most commonly used computer programs for performing factor analysis automatically rotate the factor solution to remove any general factor, and attempt to impose the requirement of at least one (near)-zero loading in each row and column of the factor matrix. The net effect of this is to provide factor solutions which suggest the existence of mental faculties along the lines of Thurstone's Primary Mental Abilities. So it is solemnly stated that a particular child, say, is good at tasks involving verbal skills because of superior verbal ability, and look!—here is a factor analysis that demonstrates the reality of verbal ability.

Once Thurstone's approach became dominant, the so-called British school was seen as outdated and was progressively ignored in the technical factor analytic literature. One outcome has been that there has been little real progress in the study of individual differences in ability from a psychometric angle for decades.

An innovation with major implications for the development of factor analysis was Hotelling's (1933) development of an efficient and practicable method for calculating the latent roots and vectors of a correlation matrix (otherwise known as eigenvalues and eigenvectors). It was the lack of such a method which made Pearson's (1901) original suggestions concerning the redescription of correlational data a non-starter for practical factor analysis at the time. These days, obtaining the eigenstructure of a correlation matrix in a fraction of a second is no problem even on quite modest computer hardware. In the 1930s it was laborious even with Hotelling's algorithm.

These developments led to a new round of acrimonious debate, particularly among Thurstone, Hotelling, and Kelley. The hot topic was what values to place in the diagonal of the correlation matrix before extracting factors. Hotelling left the obvious value of 1.0 in place, and his method became known as 'principal components'. It is now used when the point of the analysis is simply to reproduce the original scores as nearly as possible from a reduced number of dimensions. The alternative ('principal factors' or, confusingly, 'principal axes') is to put some reduced value in the diagonal, such as the reliability of the test concerned, its squared multiple correlation with all other tests in the set being analysed, or an estimate of the square of its correlation with the set of factors to be extracted (the 'communality').

This debate is recognizably modern. Burt was not a party to it, or at least he did not adopt a partisan position. He repeatedly attempted to show how the various methods and positions adopted were related to one another, and could be seen as laying different emphases rather than being strictly in conflict.

By 1940 and the publication of his textbook, he appears content to recognize that others have gone well beyond his own work, provided mathematically superior accounts of factor analysis, and indeed he recommends their textbooks. I have the impression that while by this time he has mastered much more in the way of mathematics than he had in his early career, he is not confident that his abilities in this are a match for those of transatlantic researchers. In 1940, for instance, he claims to have extracted 10 factors from a 6×6 correlation matrix. But in his 1947 review of Thurstone's *Multiple factor analysis* he recognizes that the number of factors is limited by the rank of the matrix.* He seems to be following developments and not even trying to lead.

Two cultures

It is worth drawing some contrasts between Burt and Thurstone, not least because it seems to me that the contrasts Thurstone himself wanted to draw do not really capture the difference between the British approach to

---

* The rank of a matrix is always less than or equal to its order; the order of a correlation matrix is the number of rows or columns.

factor analysis (and particularly Burt's) and the American approach, which in its Thurstonian variant became dominant. Burt began as a classicist, and as we have seen brought some classical intellectual paraphernalia with him to factor analysis. Thurstone began as an engineer: he worked with concrete geometrical shapes in his exposition of factor analytic techniques. He was interested in proofs and exactness. One of Thurstone's achievements was to bring to widespread attention what could *not* be demonstrated by factor analysis. In particular, he showed that two tests alone cannot determine a common factor. His arguments for 'simple structure' and the criteria to be applied in its pursuit were also based on the need to overdetermine, in a mathematical sense, the reference axes for rotation of factors. This is factor analysis wearing its scientific hat, although to be truthful it has to be said that many applications of a Thurstonian style of analysis have failed to take account of the implications of his findings. He worked with 56 tests, enough to overdetermine his multivariate space. For many subsequent researchers, simple structure has become a matter left to the discretion of computer programmers whose products they have bought, its precise definition having been lost to collective memory.

Reduced to a verbal account, Thurstone's criteria require well-delineated contrasts to be present in a correlation matrix to assure the identification of factors. The number and kind of variables chosen determine the extent to which this will be achieved. The dispute between adherents of *g* theory and Thurstone as to the reality or otherwise of a general factor was partly defused by the recognition that Thurstone's primary mental abilities were positively intercorrelated, and could themselves be factored to yield a second- or third-order *g* factor. The debate has not gone away, but it will never be resolved on factor-analytic grounds alone.

On the other hand, Burt's approach was to develop an art of factor analysis. His methods are not easily reduced to a computer program because they involve judgements at every stage, and as he describes them some of these judgements are best made on the basis of experience. In its mature phase, his was a four-factor theory, in the sense that he admitted factors at four levels, corresponding to the scheme of predicables mentioned above: a general factor, group factors, factors corresponding to the unique but reliable parts of individual tests, and random factors varying from occasion to occasion. One may recognize that all of these may exist, and that a skillful practitioner using Burt's methods might disentangle them. But the methods of themselves do not guarantee their discovery, and alternative solutions are always possible. Burt's criteria for meaningfulness rested on the fit of results to other facts or beliefs outside the analysis, whereas Thurstone's were internal to his methods. If such a description matters, Thurstone's approach is more mathematically objective, but it has not been any the more successful for all its scientific credentials. His 'primary mental abilities' have not heralded a new dawn of understanding of cognitive functioning, rather they seem from this distance in time as principles of classification of types of test item.

This is where I think Burt's distinctive contribution in his mature phase of factor analytic work has been overlooked, or occasionally maligned. Although *The factors of the mind* contains a lot of outdated algebra, and his pursuit of factors derived from correlations amongst people was a blind alley, his discussion of the status of factors is subtle and instructive even today (Burt 1940, p. 13):

In my view the *primary* object of factorial methods is neither causal interpretation, nor statistical prediction, but exact and systematic description. And I suspect that most of the confusion has arisen because factors, like the correlation coefficients on which they are based, have been invoked to fulfill these three very different purposes, and so have made their appearance at three very different levels of thought—like the famous legal firm of Arkles, Arkles & Arkles, which, 'more to its own satisfaction that that of its clients, canvassed three different lines of business in three small offices on three different floors'.

Throughout the text Burt maintains the tension between these three aspects of factor theory, insisting that at bottom factors express distinctions that can be drawn based on differential responses of a group of people to a battery of tests, but recognizing that for practical purposes the language of causality is difficult to avoid. It is a tension that will be familiar to any applied scientist attempting to maintain an appropriate degree of rigour and detachment in his science, whilst needing to communicate and to practise effectively in his field of application. Burt has been criticized for this stance: in other circles it is called 'tolerance of ambiguity' and is accounted a virtue.

He grasped, in a way that is rarely evident in psychology, that analogies with physical sciences should be handled like for like, that abstruse and abstract theory may have to be simplified and made concrete for practical or pedagogical purposes, and that the physical scientists who are held up as examples of how to do real science are characters in a fairy tale told to schoolchildren.

Contrary to what others have thought, I do not find Burt's work after 1940 particularly original or forward looking in the development of factor analysis. Indeed from the late 1930s onwards he seems mostly to have been engaged in showing the relationships amongst different methods, elucidating the properties of particular methods, or synthesizing the work of others. Others had overtaken him, were better equipped mathematically, and, of course, were younger. What Hearnshaw terms the 'new story' concerning the origins of factor analysis was, I think, in part an attempt to show that his exclusion from the received view was unjustified. Burt was right: it quite clearly was unjustified.

In promulgating the story, emphasizing the priority of Pearson, and laying stress on his own contributions, he was not totally accurate. As we have seen, Pearson did lay the groundwork for modern factor analysis, and Burt may well have been inspired by Pearson in the development of his own methods, but it did not look clear cut at the time. There are also, leaving aside the provenance of Appendix V to Memorandum II of *Mental and scholastic tests*, inaccuracies in Burt's references to his early work. For instance, a footnote in the fourth edition of *Mental and scholastic tests* claims that Burt described his methods in a British Association paper in 1910 (Burt 1910). He did not—or at least they are not to be found in the printed summary. Nor are they to be found in Burt (1911) as he claimed in the same footnote. But it is also clear that he is not just looking to grab glory for himself. In attributing the formulation of multiple factor analysis to Garnett he is identifying a line of thought going back to Pearson which was a much earlier statement from which the American 'rediscoverers' of the 1930s could have learned and which they should, in line with the usual proprieties, have acknowledged.

Some conclusions

The history of the development of factor analysis is not edifying. Pearson and Spearman, who together had the skills to develop it rapidly, quarrelled and feuded. Spearman stuck stubbornly to his tetrad difference equation when it was long obsolete, and his methods were simply abandoned in time. To be a little fanciful, one might see some kind of a link with modern analysis of covariance structure methods, but in so doing one would run the risk of being accused of letting sentiment run ahead of reason.

Thomson, whose work had an originality and sparkle that is often missing in the field, has been neglected and is remembered now as a critic of Spearman rather than an original thinker in his own right. Garnett, whose name has passed from the record in a way I find extraordinary, was ignored for all practical purposes. The new wave of American factorists in the 1930s lumped all the British workers in the field together, called them Spearman's school, and set out to rubbish them. Thurstone in particular turned the clock back in terms of the substantive nature of his theory, and failed to acknowledge priorities whilst claiming credit for himself. The squabbles of the 1930s have continued with scarcely a break: perhaps it is so in all disciplines, but the once rumoured threat of one prominent figure to sue a respected journal for 'knowingly publishing false algebra' gives the tone of the unseemly knockabout which, one must suppose, nurtures the advance of knowledge.

Burt, for all that he was the first to develop a practical method of factor analysis and apply it widely, to considerable effect from the point of view of public policy and educational practice, never moved far beyond his original method and was overtaken by others who never acknowledged his contribution. His real influence is felt through the work of those of his students, such as R. B. Cattell and H. J. Eysenck, who went on to build factorial theories that have attracted widespread attention, and through the continuing line of thinking which is prepared to allow both g and more particular kinds of ability continuing rôles in accounting for differences in intellectual performance. In direct line of succession one thinks of the work of Philip Vernon, and more recently of course Jensen has laid emphasis on the value in Burt's contributions. The work of Humphreys (1985) also, to some extent, keeps alive Burt's style of thinking.

So in his claims concerning his own part in the development of factor analysis was Burt lying? exaggerating? purveying unalloyed truth? I think the truth goes something like this. In the first years of his career, Burt was trying to develop an independent line of work, but as surely as any young man in a hurry he also needed the support of his seniors. The early focus of interest was on g rather than on factor analysis, and Burt's mathematical skills were not very highly developed. That his style of analysis survived and was widely applied for practical purposes was the result of his appointment to the London County Council. Until he was secure there, he could not afford to make too many waves, however independent his thinking. He may not, at the time, have realized the importance of what he was doing. He may even have been totally mistaken as to what he was about, only recognized it when Thurstone published, and felt foolish for having been unnecessarily deferential. Forty years later, he remembered his independence of mind and could trace a line of development that would not have been clear at the time. For that matter, Spearman and Thomson and others would not, at the time, have been able to discern which way factor analysis was going to go. But by now his influence, much of which could be traced to the success of studies using methods he had devised and no one else had claimed as their own, had been considerable in Great Britain.

In applying his mind to the practical problems he set out to tackle, he had neglected to publish in the right journals at the right time (like most practitioners, myself included), supposing, not unreasonably, that publication in book form of his empirical findings together with notes on method was enough to establish the extent of his contribution. It was only with the emergence of the American school of factor analysts in the 1930s that he began to realize that he was going to be written out of history, and by that time he had ceased to produce original methodological contributions. After the war, the opportunity arose to try to set the record straight, and with *amour propre* dented by what Thurstone had written, he began to see a clearer pattern in his early work. By this time, Pearson's original mathematics would have been accessible to him, and he could reassert an intellectual lineage to which he was emotionally attached, there no longer being any need to placate his elders and betters.

This is the story of a man of ordinary academic vanity, shading his own history rather more to his favour than is perhaps justified by the facts, and seeking to redress a perceived injustice suffered at the hands of a newly dominant competing school of thought. More than his own reputation, he seems to have cared for the Galton–Pearson–Garnett line of thinking, which was undoubtedly real, though having little impact in the early years on actual practice other than through his own activities. He himself, through his first publication and his subsequent adherence to the reality of g, could easily be seen as belonging to Spearman's school by those who chose not to look too closely. Spearman stood between him and the reputation he believed he deserved. It is a fact that Burt's methods and findings in factor analysis have had far more practical impact than Spearman's, and that much of Spearman's mature phase of work, in particular his enunciation of various laws concerning g, is now all but forgotten.

I can find nothing in what Burt wrote to support Hearnshaw's characterization of his behaviour as delusional. Rather, I think he showed considerable tolerance in his reviews of Thurstone's work, and some considerable skill and wisdom in later life pointing out how various methods of factor analysis and indeed analysis of variance cohere rather than compete. To the extent that he exaggerated his own importance, and I think that extent is small, and possibly manufactured evidence to secure his claims, he was foolishly trying to gild an otherwise perfectly serviceable dandelion.

As to credit for the origination of factor analysis, my reading is as follows. Pearson introduced the idea of analysing correlations in terms of principal axes, and his work is the foundation of all modern computational methods in factor analysis. Spearman introduced the notion of a factor as a psychological principle, and developed his own method of analysis which proved cumbersome and was eventually abandoned. Burt developed a simplified approximate method which was practical given the technology of the time, and he was first in the active pursuit of group factors looking at residual correlations after a general factor had been extracted. Thomson challenged Spearman's interpretation of his findings in a way that led Garnett to provide a sound mathematical formulation of factor analysis. These five together, and no one of them singly, were the creators of the factorial approach to individual differences.

This is a mixed verdict, and in reaching it by rereading the early sources I have come to a view of Burt that is even more complex than before. I still admire the energy and the enterprise in being the first to tackle a raft of complex problems with a certain degree of methodological rigour, and the later subtlety of thought and overarching view of the range of techniques flourishing in mid-century. But there is an evasiveness, a tendency to gloss over detail, a lack of frankness in description which is difficult to pin down or tie to a particular passage in his work, which leaves a sense of unease. The history of Appendix V to Memorandum II in *Mental and scholastic tests* is at the very least disturbing, but now, I think, can only be settled by a detailed critical examination of internal evidence. It may be just that Burt thought himself the intellectual inferior of such as Pearson, Spearman, and Garnett, and skimmed over detail and difficult issues in a way which flavours his prose. But I suspect that there is more to discover, and that the controversy over his early work is far from over.

Acknowledgements

Numerous people have been generous with their time and attention during the preparation of this chapter, in particular: Christine Anderson and Susan Tarrant at the University of London Library and Lyn Naylor of the University of Liverpool archives greatly facilitated my access to various sources; Gavin Ross of Rothamsted Experimental Station, Joe Marsh, honorary archivist at the University of Manchester Institute of Science and Technology, Mrs Peggy Jay (Maxwell Garnett's daughter) and Professor Julian Hunt FRS (his grandson) provided the clues that led me to recognize Garnett's contribution; and Nick Mackintosh not only showed great tolerance as editor, he actively collaborated in finding material, and first spotted that the first edition of Mental and scholastic tests did not contain the crucial appendix.

Remaining errors and omissions are of course my own.

References

Bickersteth, M. and Burt, C. (1916). Some results of mental and scholastic tests. *Report of the fourth annual conference of educational associations held at the University of London*, pp. 30–7. A copy of this paper is in the Burt archive.

Boring, E. G. (1950). *A history of experimental psychology*. Appleton Century Crofts, New York.

Brown, W. (1910). Some experimental results in the correlation of mental abilities. *British Journal of Psychology*, **3**, 296–322.

Burt, C. (1909). Experimental tests of general intelligence. *British Journal of Psychology*, **3**, 94–177.

Burt, C. (1910). Experimental tests of general intelligence. *British Association Annual Reports*, **79**, 804.

Burt, C. (1911). The experimental study of general intelligence. *Child Study*, **4**, 33–45, 92–101.

Burt, C. (1915). General and specific factors underlying the primary emotions. *British Association Annual Reports*, **84**, 694–6.

Burt, C. (1917). *The distribution and relations of educational abilities*. London County Council.

Burt, C. (1921). *Mental and scholastic tests*. King & Son, London.

Burt, C. (1937). Methods of factor-analysis with and without successive approximation. *British Journal of Educational Psychology*, **7**, 172–95.

Burt, C. (1940). *The factors of the mind*. University of London Press.

Burt, C. (1947). Critical notice of Thurstone (1947). *British Journal of Educational Psychology*, **17**, 163–9.

Burt, C. (1952). An autobiographical study. In *History of psychology in autobiography*, Vol. 4 (ed. E. G. Boring, H. S. Langfield, H. Werner, and R. M. Yerkes), pp. 53–73. Clark University Press, Worcester, MA.

Carey, N. (1916). Factors in the mental processes of school children III. *British Journal of Psychology*, **8**, 170–82.

Cronbach, L. J. (1979). Review of Hearnshaw (1979). *Science*, **206**, 1392–4.

Dodd, S. C. (1928). The theory of factors. *Psychological Review*, **35**, 211–34, 261–79.

Fancher, R. (1985). *The intelligence men*. Norton, New York.

Freeman, F. N. (1939). *Mental Tests*. Houghton Mifflin, Boston, MA.

Garnett, J. C. M. (1919a). On certain independent factors in mental measurement. *Proceedings of the Royal Society, Series A*, **96**, 102–5.

Garnett, J. C. M. (1919b). General ability, cleverness and purpose. *British Journal of Psychology*, **9**, 345–66.

Garnett, J. C. M. (1920). The single general factor in dissimilar mental measurements. *British Journal of Psychology*, **10**, 242–58.

Garnett, J. C. M. (1921). *Education and World Citizenship*. Cambridge University Press.

Gould, S. J. (1981). *The mismeasure of man*. Norton, New York.

Guilford, J. P. and Hoepfner, R. (1971). *The analysis of intelligence*. McGraw-Hill, New York.

Guttman, L. (1954). A new approach to factor analysis: the radex. In *Mathematical thinking in the social sciences* (ed. P. F. Lazarsfeld). Free Press, Glencoe, IL.

Guttman, L. (1966). Order analysis of correlation matrices. In *Handbook of multivariate experimental psychology* (ed. R. B. Cattell), Chapter 14. Rand McNally, Chicago, IL.

Harman, H. H. (1960). *Modern factor analysis*. University of Chicago Press.

Hearnshaw, L. S. (1979). *Cyril Burt: psychologist*. Hodder & Stoughton, London.

Hotelling, H. (1933). Analysis of a complex of statistical variables into principal components. *Journal of Educational Psychology*, **24**, 417–441.

Humphreys, L. G. (1985). General intelligence. In *Handbook of intelligence: theories, measurements and applications* (ed. B. B. Wolman), pp. 201–24. Wiley, New York.

Joynson, R. B. (1989). *The Burt affair*. Routledge, London.

Krüger, F. and Spearman, C. (1906). Die Korrelation zwischen verschiedenen geistigen Leistungsfähigkeiten. *Zeitschrift für Psychologie*, **44**, 50–114.

Krzanowski, W. J. (1988). *Principles of multivariate analysis: a user's perspective*. Clarendon Press, Oxford.

Macdonell, W. R. (1901). On criminal anthropometry and the identification of criminals. *Biometrika*, **1**, 177–227.

Nunally, J. C. (1978). *Psychometric theory*. McGraw-Hill, New York.

Pearson, K. (1901). On lines and planes of closest fit to a system of points in space. *Philosophical Magazine*, **2** (series 6), 559–72.

Spearman, C. E. (1904a). The proof and measurement of association between two things. *American Journal of Psychology*, **15**, 72–101.

Spearman, C. E. (1904b). General intelligence: objectively measured and determined. *American Journal of Psychology*, **15**, 201–99.

Spearman, C. E. (1928). Pearson's contribution to the theory of two factors. *British Journal of Psychology*, **19**, 95–101.

Spearman, C. E. (1930). "G" and after—a school to end schools. In *Psychologies of 1930* (ed. C. Murchison), pp. 339–66. Clark University Press, Worcester, MA.

Spearman, C. E. and Hart, B. E. (1912). General ability, its existence and nature. *British Journal of Psychology*, **5**, 51–84.

Steinberg, H. (1951). List of publications by Sir Cyril Burt. *British Journal of Educational Psychology*, **21**, 53–62.

Stephenson, W. (1983). Cyril Burt and the special place examination. *Journal of the Association of Educational Psychologists*, **6**, 46–53.

Stephenson, W. (1984). In *Sir Cyril Burt, psychologist and "the essential man": reactions by prominent psychologists to previously published accounts* (ed. C. McLoughlin). Published informally by Kent State University, and abstracted in *Psychological Documents* 14(2), 19. A copy of this is held at the University of London Library.

Thomson, G. H. (1916). A hierarchy without a general factor. *British Journal of Psychology*, **8**, 271–81.

Thomson, G. H. (1935). On complete families of correlation coefficients and their tendency to zero tetrad differences: including a statement of the sampling theory of abilities. *British Journal of Psychology*, **26**, 63–92.

Thurstone, L. L. (1924). *The nature of intelligence*. Kegan Paul, Trench, Truber, London.

Thurstone, L. L. (1925). A method of scaling psychological and educational tests. *Journal of Educational Psychology*, **16**, 433–51.

Thurstone, L. L. (1928). The absolute zero in intelligence measurement. *Psychological Review*, **35**, 175–97.

Thurstone, L. L. (1931). Multiple factor analysis. *Psychological Review*, **38**, 406–27.

Thurstone, L. L. (1935). *The vectors of the mind*. University of Chicago Press.

Thurstone, L. L. (1947). *Multiple factor analysis*. University of Chicago Press.

Thurstone, L. L. (1952). Autobiography. In *A history of psychology in autobiography*, Vol. IV (ed. E. G. Boring, H. S. Longfeld, H. S. Werner, and R. M. Yerkes), pp. 295–321. Clark University Press, Worcester, MA.

Walker, H. M. (1929). *Studies in the history of statistical method*. Williams and Wilkins, Baltimore, MD.

Webb, E. (1915). Character and intelligence. *British Journal of Psychology Monograph Supplement*, **1**, 3.

Wissler, C. (with J. M. Cattell and R. C. Farrand) (1901). The correlation of mental and physical tests. *Psychological Review Monograph Supplement* **III**.

Wolfle, D. (1940). Factor analysis to 1940. *Psychometric Monographs* **3**. University of Chicago Press.

THREE

Twins and other kinship studies

N. J. MACKINTOSH

Introduction

BURT'S PAPER 'The genetic determination of differences in intelligence: a study of monozygotic twins reared together and apart', published in 1966, provided the most systematic account of his data on kinship correlations for IQ, including what was at the time much the largest sample of separated identical (MZ) twins ever reported. Earlier papers had mentioned these data in passing, in the context of other arguments. The 1966 paper was the first whose central purpose was to present these data in their entirety with new, larger samples, to provide information about how they had been collected and analysed, and in effect to summarize his life's work on this issue. Although Burt continued to publish further papers on the heritability of IQ, he published no new data. The 1966 paper may be taken as his final, magisterial summary. At the same time, a reading of the paper makes it rather clear that it was written, at least in part, as a rebuttal of the arguments of various armchair critics who claimed, according to Burt in the absence of any evidence, that environmental factors were largely responsible for individual differences in IQ. Burt took evident pleasure in presenting data to refute their position. As if in revenge, his twin data have provided Burt's later critics with the basis for some of their most dramatic, and widely publicized, accusations of fraud.

The first charge was that although, over the years, the number of separated MZ twins in Burt's sample more than doubled, the correlations in their IQ scores often remained constant to the third decimal place (this was equally true for some other kinship categories). This was first noticed by Leon Kamin, who publicized his observations in several lectures given during 1972, and wrote to Arthur Jensen in that year to draw his attention to this remarkable feature of Burt's data. Acknowledging Kamin's role, Jensen himself published a systematic survey of all Burt's kinship correlations for IQ, and reported that he had found some 20 correlations that remained invariant in spite of apparent changes in sample size. Twenty such invariant correlations, as Jensen noted,

unduly strain the laws of chance and can only mean error, at least in some of the cases. But error there must surely be. (Jensen 1974, p. 24)

Table 3.1 Kamin's presentation of Burt's invariant correlations for MZ twins, 'group test' of intelligence

	MZa twins	MZt twins
Burt (1955)	0.771 ($N=21$)	0.944 ($N=83$)
Burt (1958)	0.771 (N 'over 30')	0.944 ($N=?$)
Conway (1958)	0.778 ($N=42$)	0.936 ($N=?$)
Burt (1966)	0.771 ($N=53$)	0.944 ($N=95$)

Kamin's own rather less systematic analysis was published in the same year (Kamin 1974). His favourite example, repeated on more than one occasion, is shown in Table 3.1. Between 1955 and 1966, Burt and his assistant Miss Conway (who was always credited with collecting most of the twin data) published four papers reporting a steady increase in the sample of separated MZ twins, from 21 pairs in 1955 to the final total of 53 pairs in 1966. Over the same period, there was a smaller increase in the size of the sample of MZ twins brought up together, from 83 to 95. As can be seen in Table 3.1, however, this steady increase in the size of the samples was not accompanied by the changes in the values of the correlations one might have expected. Kamin commented sarcastically on the 'remarkable stability' of this

unknown 'group test' of intelligence . . . There is a minor perturbation which simultaneously afflicted both correlations in late 1958, but a benign Providence appears to have smiled on Professor Burt's labors. When he concluded his work in 1966, his three decimal place correlations were back to where they had been in the beginning. (Kamin 1974, p. 38)

Kamin's conclusion was more forthright than Jensen's:

The numbers left behind by Professor Burt are simply not worthy of our current scientific attention. (Kamin 1974, p. 47)

But, although finding much else to criticize in Burt's data, he stopped well short of suggesting that any of them were fabricated. That charge was not made in public until 1976, when Oliver Gillie, writing in *The Sunday Times*, stated:

The most sensational charge of scientific fraud this century is being levelled against the late Sir Cyril Burt, father of British educational psychology. Leading scientists are convinced that Burt published false data and invented crucial facts to support his controversial theory that intelligence is largely inherited.

The scientists to whom Gillie was referring were Ann and Alan Clarke, of the University of Hull, and Jack Tizard of the University of London Institute of Education. The Clarkes had themselves earlier commented on various 'puzzling features' of Burt's data, some of which were 'suspiciously perfect' (Clarke and Clarke 1974, Chapter 7) and, according to their later testimony, although at the time too busy to pursue the matter further, had become suspicious of much of Burt's later work. But the critical new issue was that raised by Tizard who, in an attempt to get in touch with Miss Howard (one of Burt's two main assistants, with whom, in the 1950s, he published several papers), had contacted the British Psychological Society, but been informed that she had never been a member; further enquiries revealed that she had been neither a student nor a member of staff at University College since the war, and was quite unknown to a variety of Burt's London colleagues. Gillie pursued this issue of the missing assistants, found no trace of either Miss Howard or Miss Conway at University College, put an advertisement for them in *The Times* for 16 October and, getting no reply, published his article in *The Sunday Times* the following week.

It is hardly surprising that Gillie's article should have created a considerable stir. Between 1955 and 1966, Burt had apparently discovered and tested 32 additional pairs of separated MZ twins—nearly as many as the total number of twins in the largest other study. In 1955 he was 72 years old. It hardly seemed probable that he had done this work himself—indeed the published reports, one of them apparently authored by her alone, always made it clear that the bulk of the testing had been undertaken by Miss Conway. But now it appeared that she could not be found. Although many psychologists, including Jensen and Eysenck, continued to defend Burt against the charge of fraud, and although the charge had clearly not been proved, no one could deny that there were reasonable grounds for suspicion or, at the least, that there was a mystery to be explained.

The mystery was apparently solved by Leslie Hearnshaw. Although by 1979 more than one person had come forward to identify Miss Howard, there was still no evidence for the existence of Miss Conway. Hearnshaw concluded that Burt himself had written the papers published under her name—a conclusion confirmed by Burt's housekeeper and secretary, Miss

Archer, who stated that Burt had informed her that Conway at least had emigrated before 1950 (Archer 1983). It was hard to see how either could have found and tested the additional 32 pairs of separated MZ twins between 1955 and 1966, and neither his diaries nor his correspondence revealed any evidence that Burt had been in touch with either of them at any point during these years. Indeed, the diaries and correspondence yielded no mention of the testing of MZ twins or any other kinship group during this period. In the face of all this, Hearnshaw seemed entirely justified in concluding that Burt did not collect any new data after 1955, and that

he simply did not possess detailed data, at any rate for the whole sample of his separated MZ twins. (Hearnshaw 1979, p. 247)

Hearnshaw's conclusion seemed inescapable and, until the publication of Joynson's and Fletcher's books, it was almost (but not quite!) universally accepted.

What, then, is the current status of these charges and counter-charges? Are we in any position to understand why so many of Burt's correlations remained identical in spite of apparent changes in the size of his samples? Did the assistants exist, and did they collect the data Burt reported? If so, when? Although much is likely to remain uncertain and conjectural, I believe that some answers to at least some of these questions are more probable than others. Let us begin with the first question initially raised by Kamin—the invariant correlations.

The invariant correlations

The most useful analysis of Burt's kinship correlations is that provided by Joynson (1989). As we have seen, Kamin's own account was quite incomplete—although more than sufficient for the purpose of raising a serious question mark over Burt's data. Jensen's analysis was very much more systematic but, reasonably enough at the time, omitted papers published solely under Conway's name. Moreover, Jensen organized his analysis by kinship categories rather than by published papers, and this does not bring out so clearly, what Joynson unequivocally demonstrates, which papers contain the invariant correlations. The answer is that there is no mystery, and no inexplicably invariant correlations, until Burt (1966).

The first mention in Burt's writings of IQ correlations for separated MZ twins and other kinship data was in 1943, where he reported a correlation of 0.77 for the IQ scores of 15 pairs of separated MZ twins (Burt 1943). Beginning in 1955, Burt published his data rather more systematically. Table 3.2 reproduces a table of correlations from that paper. As can be seen, there were six different kinship categories, and for each category there were 11 measures, three for IQ, three for educational attainment, and

Table 3.2 Kinship correlations: Burt (1955)

	MZt	MZa	DZt	Sib t	Sib a	Unrelated t
Intelligence						
Group test	0.944	0.771	0.542	0.515	0.441	0.281
Individual test	0.921	0.843	0.526	0.491	0.463	0.252
Final assessment	0.925	0.876	0.551	0.538	0.517	0.269
Educational						
Reading/spelling	0.944	0.647	0.915	0.853	0.490	0.548
Arithmetic	0.862	0.723	0.748	0.769	0.563	0.476
General	0.898	0.681	0.831	0.814	0.526	0.535
Physical						
Height	0.957	0.951	0.472	0.503	0.536	−0.069
Weight	0.932	0.897	0.586	0.568	0.427	0.243
Head length	0.963	0.959	0.495	0.481	0.536	0.116
Head breadth	0.978	0.962	0.541	0.507	0.472	0.082
Eye colour	1.000	1.000	0.516	0.553	0.504	0.104

Mzt, monozygotic twins brought up together; Mza, monozygotic twins brought up apart; DZt, dizygotic twins brought up together; Sib t, siblings brought up together; Sib a, siblings brought up apart; Unrelated t, unrelated children brought up together.

five for various physical measures, giving a total of 66 correlations. Similar tables were published in Burt (1958), Conway (1958), and finally in Burt (1966).* Kamin, it will be recalled, noticed that the correlation for the group test of intelligence remained the same for MZ twins, both those brought up together and those reared apart (hereafter MZt and MZa), both in Burt (1958) and Burt (1966), but that both changed slightly in Conway (1958). This is correct. What Kamin fails to mention is that the table published in Burt (1958) differs from that published in Burt (1955) only in omitting the correlations for the five physical measures. It is otherwise identical: *all* six correlations for IQ and educational attainment for *all* six kinship categories in Burt (1955) and Burt (1958) are identical to the third decimal place. It is clear, and can hardly have failed to escape Kamin's attention, that Burt was simply reproducing the earlier table for this new paper (which was in fact the published version of a lecture). The only mystery about these invariant correlations is why Kamin chose to comment on them. But it would be wrong to blame only Kamin here. Burt, as often, was less than meticulous in his account of what he had done. In the 1958 paper, he wrote, *à propos* of the MZa twins

we have now collected over 30 such cases (Burt 1943, 1955). I reproduce the more important correlations for the twins in Table 1 and have added for comparison

* In fact the table published in Burt (1966) was reproduced in one or two later papers, but since no new data were added these need not concern us.

Table 3.3 Kinship correlations: Conway (1958)

	MZt	MZa	DZt	Sib t	Sib a	Unrelated t
Intelligence						
Group test	0.936	0.778	**0.542**	**0.515**	**0.441**	**0.281**
Individual test	0.919	0.846	**0.526**	**0.491**	**0.463**	**0.252**
Final assessment	0.928	0.881	**0.551**	**0.538**	**0.517**	**0.269**
Educational						
Reading/spelling	0.943	0.645	**0.915**	**0.853**	**0.490**	**0.548**
Arithmetic	0.870	0.726	**0.748**	**0.769**	**0.563**	**0.476**
General	0.894	0.629	**0.831**	**0.814**	**0.526**	**0.535**
Physical						
Height	0.956	0.942	**0.472**	**0.503**	**0.536**	**−0.069**
Weight	0.929	0.884	**0.586**	**0.568**	**0.427**	**0.243**
Head length	0.961	0.958	**0.495**	**0.481**	**0.536**	**0.116**
Head breadth	0.977	0.960	**0.541**	**0.507**	**0.472**	**0.082**
Eye colour	1.000	1.000	**0.516**	**0.553**	**0.504**	**0.104**

Numbers in bold are identical with those reported by Burt (1955).

corresponding coefficients obtained from other pairs, both related and unrelated. (Burt 1958, p. 7)

No one reading that sentence, and without Burt's 1955 table beside them for comparison, could have suspected that the 1958 table was *not* based on the new, larger sample of MZa twins.

The next paper in the series is Conway (1958). Table 3.3 reproduces her relevant Table 1. Once again, there is no mystery. All the MZ correlations, both MZt and MZa (except those for eye colour), have changed from those published by Burt (1955). All the other correlations, for all other kinship categories, remain exactly the same. Conway (1958) explicitly states that the sample of MZa twins has increased to 42 pairs, so it is unsurprising that all MZa correlations should have changed. There is no mention of the new sample size for any other kinship category, but it is reasonable to assume that there has been some increase in the number of MZt twins, and obvious that there has been no change in any other category.

Thus there are no inexplicable, invariable correlations up to and including Conway (1958). They are, in fact, all confined to the final paper in the series, Burt (1966). Table 3.4 reproduces the relevant table from that paper, with the same convention as I used for Table 3.3 above: correlations repeated from either Burt (1955) or Conway (1958) are printed in bold. If we ignore the two MZ correlations for eye colour (since although repeated, the correlation of 1.00 is hardly mysterious), we are left with 31 of the remaining correlations that are identical, to the third decimal place, with those published earlier.

Table 3.4 Kinship correlations: Burt (1966)

	MZt	MZa	DZt	Sib t	Sib a	Unrelated t
Intelligence						
Group test	**0.944?**	**0.771**	0.552	0.545	0.412	**0.281**
Individual test	0.918	0.863	0.527	0.498	0.423	**0.252**
Final assessment	**0.925**	**0.874**	0.453?	0.531	0.438	0.267
Educational						
Reading/spelling	0.951	0.597	0.919	0.842	**0.490**	0.545
Arithmetic	**0.862**	0.705	**0.748**	0.754	**0.563**	0.478
General	0.983?	0.623	**0.831**	0.803	**0.526**	0.537
Physical						
Height	0.962	0.943	**0.472**	0.501	**0.536**	−0.069
Weight	0.929	0.884	**0.586**	0.568	0.427	0.243
Head length	**0.961**	**0.958**	**0.495**	**0.481**	0.506	0.110
Head breadth	**0.977**	**0.960**	**0.541**	0.510	0.492	**0.082**
Eye colour	1.000	1.000	**0.516**	0.554	0.524	**0.104**

Numbers in bold are identical with those reported either by Burt (1955) or Conway (1958).

The three queries are against numbers about which there is some question (see text for details).

And now there really is a mystery that demands explanation. One or two invariant correlations might be expected by chance. The addition of a small number of new subjects to an already large sample might not necessarily change the value of the correlation (especially that of a high correlation whole standard error is small). But we can be quite confident that chance alone will not explain why nearly half of these correlations are identical with those published earlier. The only possible explanation is that they have been reproduced from one or other of the earlier tables. That seems obvious enough, but hardly solves the mystery. *Why* were they copied from earlier tables? And why were *these* particular correlations and not others, so copied?

There is, of course, no mystery about a repeated correlation where the sample size has not changed.* But that simple explanation is not sufficient

* Unfortunately, we do not always know the precise size of Burt's samples. The critical information is often lacking and, where provided, sometimes bizarrely inconsistent with other information. Only in the 1966 paper are the relevant numbers given in the appropriate table of correlations. Except for the MZa sample, no numbers at all are given in Burt (1958) and Conway (1958). Burt (1955) gives a set of numbers in the text, as follows: 'The total numbers now amount to 984 siblings, of whom 131 were reared apart; 172 dizygotic or two-egg twins, all reared together; . . . [and] 287 foster children' (Burt 1955, p. 167). Are these total numbers, or numbers of pairs? At any rate, the comparable numbers in 1966 are 264 sibs together, 151 sibs apart, 127 DZ twins, and 136 unrelated children together. If both sets of figures are referring to pairs, then in three of four cases the sample size has apparently decreased, and (in the light of what will become evident below about the number of errors in Burt's 1966 table) it is at least possible that the apparent changes in the numbers of sibs apart from 131 to 151 and of DZ twins from 172 to 127 might be misprints.

here. There is no kinship category where all the correlations remain the same, so that we could conclude that there had been no increase to this particular sample. There is, equally, no kinship category where none of the correlations remain the same. Finally, there is also no single measure for which all the correlations remain unchanged, so that one could argue that although the sample sizes had all changed, Burt had not collected new data for one particular measure, and so had simply repeated the old correlations. It is the very haphazard pattern of changed and unchanged correlations in Table 3.4 that creates the problem (contrast this with Table 3.3 above from Conway 1958).

Joynson (1989) agrees that where the correlations remain the same, Burt must have been simply using his old figures. He considers various possible explanations, some suggested by Banks (1983) and Cohen (1983), two of Burt's staunchest defenders, and concludes that in most cases, most probably, Burt repeated old figures because although the total size of each sample had indeed changed, he had not always collected new data on all possible measures for the newly added subjects. Since there is simply no way that a reader of the 1966 paper could be expected to infer this (without comparing the 1966 table with those published earlier, and thus noticing the invariant correlations), this hardly exonerates Burt from a charge of deception. Joynson's defence seems curious:

Why did not Burt, when writing his 1966 paper, point out that, if the reader compared the current figures with the old, he would find anomalies—some correlations remaining unchanged though the numbers [in the samples] had apparently altered?

I think the answer to this is probably that Burt did not think about it. Burt would be concentrating on the figures for intelligence. The repeated coefficients predominate in the physical and educational measures. Even if he had reflected on the number of those correlations that were repeated, he would have thought it would be obvious to the reader that he did not need any more, so had not collected any more. (Joynson 1989, p. 154)

It is true, as can be seen from Table 3.4 above, that the majority of the invariant correlations appear in the physical measurements. No doubt these are less important, and were so regarded by Burt. But it hardly follows that it was reasonable to expect his readers to see for themselves, without any help from him, that he had simply not bothered to collect new data in these cases. The defence is not helped by Fletcher's contribution here. Discussing the physical measurements, Fletcher notes:

Throughout this whole sequence of studies and articles, Burt made it quite clear that (apart from the fact that these were of less interest to him than the mental and scholastic measurements) his calculations here were only and consistently based upon *small samples*. The *number* did not vary with the increase in the total number of siblings, as did (obviously) the other tests. They were deliberately limited to a smaller size throughout. (Fletcher 1991, p. 301, italics in original)

It is true that in 1955, Burt wrote:

The figures for head-length, head-breadth, and eye colour are based on much smaller numbers in every batch. (Burt 1955, p. 167)

But in 1966, he was more explicit, and what he says is not what Fletcher implies:

Assessments of eye-colour, head-length and head-breadth were obtained for all twins, but only random samples (fifty of each sex) in the case of ordinary siblings: measurements for height and weight were corrected for sex and age. (1966, p. 141)

This is further clarified by a footnote to the table, as follows:

In columns, 3, 4, 5 and 6 the correlations for head-length, head-breadth, and eye colour were based on samples of 100 only. (p. 146)

Since column 3 is DZ twins, it is clear that by 'twins' in the earlier statement, Burt meant only MZ twins. Note that it is only three of the physical measures that were based on these less than complete samples. Taking Burt at his word, therefore, we should expect to see new correlations for the remaining two physical measures, height and weight, for DZ twins, both sets of siblings, and unrelated children reared together. In fact, seven of these eight correlations are invariant. Similarly, if we supposed that the 100 randomly chosen subjects in each of these categories came from the earlier samples and did not include anyone from the new, 1966 samples, we might expect to see the correlations for head-length, head-breadth, and eye colour for these four groups to remain the same. In fact, six of twelve are new. Finally, in the case of MZ twins, of course, all the correlations should be new, but (ignoring eye colour) six of eight are unchanged. Although Joynson and Fletcher are correct in pointing out that the majority of the invariant correlations occur in the physical measurements, the latter's account is quite insufficient to explain the nature of these invariances.

What, then, are we left with? We should certainly acknowledge one possibility (to which Fletcher pays rather more attention than Joynson). Some of the numbers in Table 3.4 are quite certainly misprints or due to other, similar types of error. We know this in two of the cases marked with a question mark, because Burt gives a different (and more plausible) value for the correlation in the text of his paper. The figure of 0.983 for general educational attainment of MZt twins is improbably high, and in the text is referred to as 0.89. The figure of 0.453 for final assessment IQ for DZ twins is referred to in the text as 0.54 (which is more in line with the figure of 0.551 in Burt (1955) and Conway (1958). Finally, according to Jensen, Burt informed him that the figure of 0.944 for the group IQ test for MZt twins was not in fact the same as the value obtained for this correlation in 1955.

Although the 1955 paper reported a value of 0.944, this was a misprint for 0.904.

There is, of course, no way of knowing how many other figures in Table 3.4 may be in error. But such errors may help to explain some of the apparently changed correlations. Perhaps Burt had meant to copy down an old correlation, but had made a mistake either in copying or in proof-reading. We can hardly appeal to this sort of error to explain many of the *invariant* correlations without postulating coincidences every bit as improbable as would be required for a newly calculated correlation actually turning out to be the same as an old one (we are already appealing to one such coincidence to explain why the MZt twin correlation of 0.944 incorrectly reported in 1955 just happened to be the same as the new correlation reported in 1966). But copying or proofreading errors might explain some of the 33 apparently new correlations in Table 3.4, and thus help to explain the otherwise inexplicable *pattern* of new and old correlations. Thus to take one example: if, in the case of DZ twins, we supposed that 0.552 (group IQ) was a misprint for 0.542, 0.527 (individual IQ) a misprint for 0.526, and 0.919 (reading and spelling) a misprint for 0.915, there would be *no* new correlations in this category.

I believe that this probably is part of the explanation of the pattern of correlations that we have in Table 3.4. But, unless we are prepared to postulate this sort of error on a truly heroic scale, it will hardly suffice. We need, I believe, to postulate a different kind of error. I suggest that when he was compiling the table for his 1966 paper, Burt had available (on various scraps of paper?) three different sets of figures: those for the 1955 paper; the new numbers for MZ twins from Conway (1958); and some new numbers for the final 1966 samples.* But because he did not sit down to do the job all at once, he was careless, forgetful, or simply got muddled, and copied down numbers now from one source, or scrap of paper, now from another.

On the face of it, this suggestion will probably seem merely ludicrous, and is hardly made more plausible by noting that Burt was 83 years old in 1966, and readily acknowledged his frailties:

What I write has to be checked and re-written many times before it is fit for the printer. Most of the mistakes are quite childish. (Hearnshaw 1979, p. 242, citing a letter from Burt to his sister)

There is, however, good independent evidence that Burt was capable of precisely this sort of error when writing his 1966 paper. The full 1966 table, like those published earlier in Burt (1955) and Conway (1958), contained not only Burt's own correlations, but also, for comparison, information

---

* It does not matter for my argument whether these new numbers were genuine, or simply fabricated; nor do I commit myself on the question whether there was a complete set of 66 new numbers or only an incomplete sub-set.

Table 3.5 Data from Newman *et al.* (1937) reported by Burt (1955), Conway (1958), and Burt (1966)

	5A Burt (1955) and Conway (1958)			5B Burt (1966)		
	MZt	MZa	DZt	MZt	MZa	DZt
Intelligence						
Group test	0.922	0.727	0.621	0.922	0.727	0.621
Individual test	0.910	0.670	0.640	0.881	0.767	0.631
Educational attainment	0.955	0.507	0.883	0.892	0.583?	0.696
Physical						
Height	0.981	0.969	0.930?	0.932	0.969	0.645
Weight	0.973	0.886	0.900	0.917	0.886	0.631
Head length	0.910	0.917	0.691	0.910	0.917	0.691
Head breadth	0.908	0.880	0.654	0.908	0.880	0.654

Of the 21 correlations reported by Burt (1966), 12 are the same as Newman *et al.*'s raw correlations, 6 are the same as Holzinger's corrected correlations, and 15 are the same as Newman *et al.*'s partial correlations. All three sources must have been used.

The ? against 0.930 (DZt, height in Table 3.5A) is because this is Burt's misprint for 0.934. The ? against 0.583 (MZa, educational attainment in Table 3.5B) is because this figure comes from a neighbouring row in Newman *et al.*'s original table.

from an earlier twin study, that of Newman, Freeman, and Holzinger (1937). This part of Burt's tables is reproduced here as Table 3.5. Table 3.5A shows the numbers as reported by Burt (1955); Table 3.5B shows those reported by Burt (1966). Conway (1958) reproduces exactly the same numbers as those given by Burt (1955)—down to and including one misprint from Newman *et al.*'s original table. As before, therefore, there is no puzzle until we come to Burt (1966)—where, as can be seen from a comparison of Tables 3.5A and 3.5B, no fewer than 10 of Newman *et al.*'s 21 correlations differ in the two versions. What can possibly have happened?

Part of the mystery is more apparent than real. In 1955, Burt was reporting Newman *et al.*'s raw correlations. In 1966, however, he wrote:

their raw figures were corrected for age and range by McNemar, and the slight changes this involves have been accepted by Holzinger. (Burt 1966, p. 145)

In a review of Newman *et al.*'s book, McNemar (1938) had noted that the variability in the MZa twins' scores was less than that of the other two groups of twins and, applying a standard correction for restriction of range, had calculated new correlations for the IQ scores. Holzinger (1938), the statistician of the original study, had replied to McNemar's review and published a new table of corrected correlations. Some, but by no means all, of the numbers reported by Burt in 1966 come from this source. But he was wholly unsystematic about this, since some of the numbers in his table are the original, uncorrected correlations. And to make matters worse, some of his numbers come from a third source—correlations with age partialled out also published by Newman *et al*. In other words, the numbers reported by Burt (1966) as Newman *et al*.'s correlations come from three different sources: some are the raw correlations (Table 96 in Newman *et al*.), some are the partial correlations (Table 25 in Newman *et al*.), and some are the correlations corrected for restriction of range published by Holzinger. In some instances, these corrections do not affect the value of the correlation, and sometimes both corrections produce the same new value. So there is no way of knowing for certain in all cases which particular source was used. What is certain, however, is that all three must have been used: no one source, or even any combination of two, is sufficient to produce the numbers shown in Table 3.5B. In fact, there is a fourth source—an error of copying. As Jensen (1974) noted, the value of 0.583 for MZa twins' general educational attainments is in fact the number from a neighbouring row (for the Woodworth–Mathews personality inventory) from Newman *et al*.'s Table 96.

Where we have an independent check on the origins of Burt's (1966) table, therefore, we can be confident that the numbers he reports come, almost at random, from three separate sources, plus one error of copying. This suggests that the most plausible explanation of that part of his table which represents his own data (Table 3.4 above) is that these numbers also come from three different sources (Burt 1955; Conway 1958; some new numbers), with some admixture of error. There is, surely, a further implication. Whatever else it may suggest, Burt's inability in 1966 to copy down a single set of numbers from Newman *et al*. is not evidence of fraud. By the same token, therefore, the rest of Burt's 1966 table is evidence of nothing worse than muddle and carelessness: it cannot be used to prove that any of the numbers were simply invented. Of course they might have been: that is to say, the third source of the numbers in Table 3.4 might not have been numbers calculated for genuinely new samples, but simply numbers invented to give the impression that he had collected new data. But if the accusation of fraud is to be upheld, it must be based on quite different arguments.

This is not really a new point. As numerous commentators have noted, if Burt had wanted to commit fraud, the one thing he would not have done is to report any invariant correlations, for as soon as Kamin pointed to such invariances everyone knew that something was wrong. Even more preposterous is the apparent implication of Gillie's statement in *The Sunday Times*:

Burt miraculously produced identical answers accurate to three decimal places from different sets of data—this is a statistical impossibility and he could have done it only by working backwards to make the observations fit his answers.

What Gillie seems to suggest here is that in 1966 Burt went to the trouble of inventing IQ scores for 'new' pairs of twins which satisfied his requirement that the correlation for the new larger sample would be exactly the same as that for the old smaller sample. Not only is it absurd to suppose that Burt would want the two correlations to be identical; Gillie seems to be suggesting that Burt *needed* to invent the IQ scores that would produce a desired correlation. But Burt did not present any IQ scores for any of his kinship correlations. He reported nothing but the correlations. If he was engaged in fraud, all he needed to do was invent some correlation coefficients.

The missing assistants: when were the data collected?

When Kamin and Jensen documented Burt's invariant correlations, neither jumped to the conclusion that Burt's data were fraudulent (although Kamin later claimed that he had intended that his readers should draw just this inference). Gillie's public accusation of fraud was largely based on his inability to find any trace of the assistants who had supposedly collected the additional data between 1955 and 1966. To Gillie (and many others) this implied that no such data had ever been collected. And Hearnshaw's further research added force to this implication by apparently showing that, whether or not the assistants had ever existed, neither Burt nor any other psychologist had been in contact with them during the critical years when 32 additional pairs of MZa twins had been added to the sample.

Gillie's *Sunday Times* article provoked much controversy in the correspondence column of *The Times*, *New Statesman*, and other newspapers and journals. Two correspondents, Professors J. Cohen from Manchester and D. Macrae from the London School of Economics, soon provided some evidence of the existence of Miss Howard, Cohen from the 1930s, and Macrae from 1950. Miss Conway proved more elusive, but Fletcher (1991, 1993) eventually traced a Miss Conway who had been a schoolteacher and student of Burt's in London during the 1930s. But if this was, indeed, Burt's Miss Conway, then Miss Archer's recollection that Burt had told her she had emigrated in 1950 was not far from the mark. She married in 1945, and settled in Dublin. As far as her children were aware, she had no further contact with Burt: indeed, according to their testimony, she had told them that she had simply been a student of Burt's before the war, rather than any form of research assistant, and they 'were sure she would have made some mention' of the fact that she had tested twins for Burt, if she had actually done so (Fletcher 1993).

In fact, the question of Miss Howard's or Miss Conway's existence is something of a red herring. However convincing the evidence that they actually existed, it would not be sufficient to answer the critical question: did they do what Burt claimed they did, namely help to collect data on the

IQ scores, educational attainments, etc. of the various kinships (most importantly, no doubt, those of the MZa twins) reported in Burt (1966)? There is, in fact, no reason to believe that they did, and very good reason to believe that, if they did, it was not after 1955. Hearnshaw, with access to Burt's papers, is categorical here:

On this issue the evidence from the diaries is decisive. Though there are some gaps in the diary record, the diaries are so nearly complete, for fifteen of the last eighteen years of Burt's life, and record so many trivia (haircuts, tea in the garden, walks on the hill, the temporary disappearance of the cat, etc.) as well as listing engagements of his own and visitors to the flat (even the weekly Saturday visits of Charlotte Banks), that we can be reasonably confident that no important activity or contact has been omitted. On the basis of this evidence, we can be sure that Burt himself did not collect any data on twins, or any other topic, during these years, and that he was never visited either by Miss Howard, or by Miss Conway, or by any other assistant actively working for him. (Hearnshaw 1979, p. 240)

Both Joynson and Fletcher have strongly disputed Hearnshaw's account, arguing that he greatly exaggerates the completeness of the diary record, and that the evidence of the diaries alone cannot be regarded as decisive. We can accept their argument. But the diaries do not provide the only evidence. As Hearnshaw also noted, there was no trace in Burt's 'carefully filed correspondence of any communication from any of these supposed assistants' (loc. cit.). Gillie (1979) wrote to some 250 of Burt's former colleagues: not one of the 100 or so who replied had ever met either Howard or Conway. Neither Miss Archer, Burt's housekeeper and secretary, nor Dr Banks, his most regular visitor, saw anything of either at Burt's flat. There is no trace at all of Miss Howard after 1950, and we have the independent evidence of her children that the only possible candidate so far unearthed for the role of Miss Conway was married in 1945 and went to live in Dublin.

Unless Burt himself, aged over 70 and not particularly well, collected the data himself, it is hard to believe that he added 32 new pairs of MZa twins to his sample after 1955. Such doubts are reinforced by one further consideration. Both Conway (1958) and Burt (1966) state that many of the more recent additions to their sample of MZa twins were obtained from personal contact (unlike the earlier twins who had turned up in the course of Burt's routine testing in schools for the London County Council). As we shall see shortly, if Burt's figures are to be believed, 10 of the 11 pairs of twins added between 1958 and 1966 came from professional families—an observation quite consistent with these statements. So why did none of these twins, or their parents, come forward in 1976 when the scandal broke and received so much publicity in the national press? It is a striking feature of the Burt affair, indeed, that only one twin has ever come forward (a Mr Hammond, who said he had been tested during the war; see Hearnshaw 1979, p. 243). To the extent that the twins were discovered during the course of routine school surveys before the Second World War, this is not surprising. But if 64 new MZa twins were discovered and tested after 1955, many of them via personal contact with friends, acquaintances, and colleagues, it verges on the incredible that there is no independent evidence for the existence of a single one.

Can we conclude, then, that the additional twins never existed? Even Hearnshaw does not draw this conclusion. He believes that Burt's MZa twin sample was never confined to the 15 pairs reported in 1943, or even to the 21 pairs reported in 1955. But he suggests that the data for the additional twins, although collected before the Second World War, were destroyed during the war when many of Burt's papers were stored at University College—which suffered a direct hit during the blitz. Embittered and frustrated in his old age, as Hearnshaw recounts, and determined to put down his armchair, sociological critics who had never themselves collected any data at all, Burt may have 're-invented' or 'reconstructed' his lost data.

This is a plausible enough story and, at first sight, seems to provide the only possible explanation of the facts we have just been considering. It is no wonder that so many of Burt's original defenders were persuaded by Hearnshaw's argument. But the new defence is simple: Burt's data were not destroyed during the war, they were temporarily lost during the course of Burt's several moves during and after the war (Banks 1983; Joynson 1989). When, after his retirement in 1951, he at last had time to analyse and write up all the data he had collected during his years as an educational psychologist, he could not immediately find them all. The increase in the sample size between 1955 and 1966 was not a consequence of the collection of new data, but of the rediscovery of old.

Is this as plausible a story as Hearnshaw's? Although those convinced of Burt's guilt will doubtless demur, it must surely be allowed that it is *possible*. But it is worth pursuing the matter a bit further. It is certainly reasonable to suppose that Burt would have been able to collect a great deal of kinship data in London schools while employed by the London County Council. But Burt's employment with the London County Council ceased in 1932. From 1930 on, the London County Council files on psychological research in schools are relatively detailed, and record lengthy discussions of many of the problems associated with such research. Although, on his move to University College, Burt established careful procedures for his students to apply for permission to do such work in London schools, the files reveal only one application between 1930 and 1940 with which Burt was associated (Sutherland and Sharp 1980). The implication is that Burt did *not* collect any further data from surveys during the 1930s, and this implication is supported by the fact that in the one paper which actually gives the dates of his surveys of London schoolchildren, the last date before the Second World War is 1930 (Burt 1969, see Chapter 5 below).

The most reasonable inference is that the large bulk of the kinship data published in 1955, 1958, and 1966 was actually collected between 1914 and 1932. If that is true, and if Fletcher's Miss Conway is indeed Burt's, then we can be confident that she did not help him in the collection of these data. She graduated with a degree in history from Trinity College Dublin in 1930, obtained a teaching diploma from King's College London in 1932, taught for some years, and was only employed by the London County Council as an assistant child care organizer between 1936 and 1945. It is only during this last period that she is likely to have assisted Burt, and it is as certain as anything can be that she did not assist him before 1930.

However, with one obvious exception, all this is by and large consistent with what Burt himself later wrote. In 1955, he made it clear that the data he was reporting had been obtained largely from 'surveys in the London schools . . . supplemented by further data collected by Miss Conway' (Burt 1955, p. 167). In 1966, he again talked of his London surveys, which had turned up a number of twins, some of whom had been separated, and Miss Conway was mentioned as having re-tested children originally tested by local teachers or doctors (Burt 1966, p. 141). The one glaring exception is that the later papers imply that new data had actually been collected after 1955:

Since the last review of our own cases was published [Burt, 1955] our collection has been further enlarged. (Conway 1958, p. 186)

And in 1966, after mentioning the 1955 and 1958 reports:

Meanwhile . . . further cases of separated twins have been brought to our notice . . . The main purpose of the present paper therefore will be to bring together the evidence now available. (Burt 1966, p. 140)

One further point, however, seems worth considering. If, as seems likely, most of the kinship data were collected before 1932, it is surely rather surprising that Burt waited so long before reporting them—over 10 years before the first brief mention in Burt (1943), another dozen years before the first moderately systematic report in Burt (1955). It is one thing to say that the war and its disruptions prevented him from reporting data collected in the late 1930s. But why should he have kept silent throughout the 1930s about data collected even earlier? At least some of the MZa twins were, supposedly, discovered and tested during the course of his London County Council surveys. Apart from a handful of isolated cases, no study of MZa twins was published until that of Newman *et al.* in 1937. Is it likely that Burt would have sat back and let the Americans not only write him out of the history of factor analysis (see Chapter 2), but also scoop his twin study? That Burt was not oblivious to such considerations is suggested by the minutes of a London County Council committee which met in 1931 to consider an application from Hogben to give intelligence tests to twins in London schools. Burt enthusiastically supported the application, regretting that 'although the problem was, in point of fact first raised in this country,' most work on twins had been done in the USA and that

no-one has hitherto investigated what Sir William Beveridge rightly calls, 'the great mass of untapped and valuable information on which light would be thrown by this inquiry in London schools'. (cited by Sutherland and Sharp 1980, p. 201)

But the later claim is that he himself had already undertaken just such an investigation.

The smoking gun

At this point, we seem to have reached an impasse. The combination of the invariant correlations and the mounting evidence that neither Burt nor his assistants had tracked down and tested 32 additional pairs of MZa twins between 1955 and 1966, seemed to lend overwhelming support to Hearnshaw's conclusion. But further reflection suggests that we should not be swayed by this combination of circumstances. There is no logical connection between them. The invariant correlations themselves do not prove fraud. We may never know their true explanation, but I believe that Burt's final table of correlations is most probably a product of error, carelessness, and sheer muddle. And we now have an alternative explanation for the 32 additional pairs of twins: they did exist, their data were recorded, but then temporarily lost and not analysed until after 1955. The prosecution may object that this is implausible; it might be on even stronger grounds if it asked why Burt waited so long to publish anything at all about his kinship data, and why he should have appeared to imply that Hogben's planned study in 1931 was the first to use the London school system as a source of twin data. But to argue that a case is implausible is not sufficient to prove fraud. It is hardly *plausible* to suppose that a widely admired and respected psychologist should, at the age of 75, start fabricating non-existent data. The prosecution is justified in complaining that Burt was guilty of concealing what he was up to. But that is a lesser charge which the defence can presumably accept.

Is there any further evidence that would help to decide between these two versions of events? There is in fact reason to believe that there is more amiss with the data presented by Burt in 1966 than we have discussed so far—and that some of these problems apply to earlier data—for example those reported by Conway (1958). On 2 December 1968, Christopher Jencks, a sociologist from Harvard, wrote to Burt asking for the raw IQ scores of his 53 pairs of MZa twins, and also for information about the socio-economic status of their homes. Normally the most punctilious of

Table 3.6 Data sent by Burt to Jencks and Jensen: final assessments on intelligence of MZa twins and social class of their homes

| Case number | Reared in | | | | Case number | Reared in | | | |
| | Own home | | Foster home | | | Own home | | Foster home | |
	IQ	Social class	IQ	Social class		IQ	Social class	IQ	Social class
1	68	6	63	6	28	97	5	92	1
2	71	4	76	5	29	97	3	95	5
3	73	5	77	5	30	97	5	112	R
4	75	6	72	R	31	97	6	113	5
5	78	3	71	6	32	99	5	105	4
6	79	3	75	5	33	100	3	88	3
7	81	5	86	5	34	101	5	115	5
8	82	2	82	R	35	102	5	104	R
9	82	4	93	2	36	103	6	106	5
10	83	4	86	6	37	105	6	109	1
11	85	5	83	6	38	106	5	107	5
12	86	6	94	6	39	106	6	108	R
13	87	5	93	1	40	107	5	108	3
14	87	6	97	6	41	107	3	101	2
15	89	6	102	2	42	108	5	95	5
16	90	2	80	2	43	111	6	98	5
17	91	3	82	2	44	112	6	116	5
18	91	2	88	5	45	114	1	104	5
19	92	5	91	6	46	114	5	125	5
20	92	2	96	3	47	115	2	108	6
21	93	6	87	2	48	116	2	116	6
22	93	3	99	4	49	118	2	116	6
23	93	5	99	3	50	121	1	118	5
24	94	6	94	R	51	125	4	128	5
25	95	6	96	5	52	129	2	117	5
26	96	2	93	R	53	131	1	132	4
27	96	4	109	1					

correspondents (many people have testified that a brief query would often yield from Burt a ten-page reply by return of post), on this occasion Burt did not reply until 25 January 1969. His reply contained the data which Jencks had requested, which are shown here as Table 3.6. Burt's covering letter began:

I apologise for not replying more promptly; but I was away for the Christmas vacation, and college (where the data are stored) was closed until the opening of term. (Hearnshaw 1979, pp. 246–7)

As Hearnshaw notes, this opening sentence contains three untruths: Burt had not been away for Christmas, his data were not stored at University College, and the college was not closed for the entire vacation. But, as both Joynson and Fletcher justly remark: which of us has not told untruths to excuse their failure to answer a letter more promptly? More damaging, however, is the entry in Burt's diary for 2 January 1969, which states that he spent a week 'calculating data on twins for Jencks', and on 11 January, 'finished checking tables for Jencks' (Hearnshaw 1979, p. 247).

The data which Burt sent to Jencks were the final assessments for IQ plus a socio-economic status rating for each of the 106 MZa twins. In 1966, he had published not only the correlation coefficient for these IQ scores, but also their mean and standard deviation; he had also presented, in that paper, a table showing the distribution of the socio-economic status ratings for the 106 homes in which the twins had been brought up (Burt 1966, Table 1). It is difficult to see how he could have done this without already having the data in some such form as that in which he sent them to Jencks. Even allowing for his age, therefore, it is difficult to see how it could have taken him a week to collect them now. But, finally, the diary entry does not say '*collecting* data', but '*calculating* data'. What was there to calculate?

Cronbach (1979, see Jensen's chapter above) referred to this entry in Burt's diary as a 'smoking gun'. It surely is. A rather plausible interpretation is that Burt was actually doing here what Gillie seemed to imply he had been doing earlier—namely inventing a set of IQ scores which would yield a desired correlation coefficient. For if he had never had 53 pairs of IQ scores, this was precisely what he now had to do: the correlation co-efficient had been published in 1966; now he must produce the scores that would yield such a coefficient. This may seem plausible, but it is a lot to hang on a single diary entry. Are there any other clues that might confirm or disconfirm such a suggestion?

Examination of the numbers Burt sent to Jencks (which were published by Jensen 1974) reveal several problems and inconsistencies. One or two of these were noticed by Kamin (1974) and Jensen (1974), and I shall not trouble to comment on them here. But there are other, more serious discrepancies. Although some may, as before, reflect error or muddle, that is not always the most reasonable interpretation. Taken together, they tend to lend weight to the suggestion of fraud.

The correlation for final assessment IQ scores for the 53 pairs of MZa twins reported in Burt (1966) was 0.874. This was stated (p. 145) to be an intra-class correlation. The intra-class correlation for the 53 pairs of scores shown in Table 3.6 is 0.877. Given that Burt's calculations were all performed on a mechanical calculating machine, his 1966 figure is close enough not to worry. More puzzling is that Burt (1966) reported the mean and standard deviation of these scores as 97.8 and 15.3 respectively, but the relevant figures for the scores shown in Table 3.6 are 97.7 and 14.9. The differences might seem too small to pay any attention to (although the calculation of a mean, at least, is rather straightforward), were it not for the fact that there is a ready explanation. In 1966 (p. 144), Burt stated that his group of MZa twins

included a couple of mental defectives from special schools (IQ 66), and, at the other end of the scale, two scholarship winners (IQ 136 and 137).

One will search in vain for such scores in Table 3.6. The two highest scores are 132 and 131. Burt's explanation of the discrepancy (in a letter to William Shockley, cited by Fletcher 1991, p. 380), is that the IQs were given as 136 and 137 by Conway (1958) but that 131 and 132 were 'our final assessment . . . in 1965'. This can hardly be true since, as we have seen, the scores of 136 and 137 were repeated in the 1966 paper.* And that these were the scores actually used in the 1966 paper is suggested by the fact that if they are substituted for 131 and 132 (and 66 is substituted for each of the two lowest scores of 68 and 63), the mean of the scores in Table 3.6 now becomes 97.8 and the standard deviation 15.1. The trouble is that these substitutions also change the correlation coefficient from 0.877 to 0.881— i.e. further away from the value of 0.874 given in Burt (1966).

None of these discrepancies is large and, given what we have already seen about the accuracy of some of the other numbers in Burt (1966), they could perhaps be dismissed as error. Further analysis, however, reveals further discrepancy between the numbers given in Table 3.6 and earlier reports—in this case Conway (1958). Conway reported a correlation of 0.881 for IQ final assessment for her 42 pairs of MZa twins. It would be a daunting task to calculate the correlation coefficients for all possible subsets of 42 pairs from the 53 pairs shown in Table 3.6. But additional information provided about Conway's sample makes it possible to identify most of the 11 pairs of twins that were added to bring the final total up to 53.

In a second paper published under Conway's name alone, written in reply to criticisms by Halsey of her earlier paper, we find the following statement about Conway's 42 pairs:

There was an appreciable number of pairs (11 of 42) in which one of the twins had been brought up in a household which Dr Halsey would certainly class as belonging to his highest category; the head of the household was a university or school teacher, a 'business owner', a 'manager', or the like . . . With none of these eleven pairs did the difference between the assessments for intelligence exceed 6 IQ points. (Conway 1959, pp. 8–9)

* It is worth pausing to ask what sort of re-assessment it is that can lop five points off two IQ scores (whether between 1958 and 1965 or later). For such an exercise to be meaningful, the two twins must surely have been retested, and Burt explicitly states that this was his standard practice (Burt 1966, p. 140). But how is this to be reconciled with the defence's only plausible position—that the twin data were collected much earlier?

Examination of Table 3.6 reveals 21 cases where one or other twin lived in a home with a socio-economic status rating of 1 or 2 (Burt's class 2 includes clergy, teachers, college lecturers etc., see Burt 1961; see also Fletcher 1991, p. 381). Since Conway states that there were only 11 such cases in her sample of 42, 10 of these 21 must be among the 11 new pairs added to bring the final sample up to 53. In eight cases, (numbers 9, 15, 16, 17, 27, 45, 47, 52), there is a difference of more than six points between the IQ scores of the two twins, so these must all be new. Conway (1959, p. 8) also states that for eight of her eleven twins living in professional households, this was their natural home, and for only three was it their adopted home. Table 3.6 reveals five additional cases where the adoptive home was of socio-economic status 1 or 2 (numbers 13, 21, 28, 37, 41). Two of these must therefore be new.

We have now narrowed down the options: we can definitely identify eight of the eleven post-Conway cases in Table 3.6; two must come from five other possibilities, and the eleventh from the remaining 40. (We also know, for what it is worth, that case 53 was definitely in Conway's sample.) It thus becomes feasible to calculate the intra-class correlations for all remaining possible combinations of 42 pairs in Table 3.6. These correlations range from 0.901 to 0.918. The correlation reported by Conway (1958) was, as we have seen, 0.881. The discrepancy is no longer so trivial, but it is also unsurprising. The correlation for all 53 cases in Table 3.6 is 0.877, but we have now had to drop eight of the cases with the largest discrepancies in IQ scores, so the correlation for the remainder is bound to be higher.

A staunch defender of Burt might insist that we have had to engage in a rather tortuous line of reasoning in order to come up with this discrepancy, and one that depends on too many possibly erroneous assumptions. There are real problems with this line of defence. If we accept, in good faith, the statements made by Conway (1959) and the numbers sent by Burt to Jencks, then it is simply not possible to reproduce Conway's correlation coefficient from any possible subset of 42 pairs out of the total set of scores shown in Table 3.6. Why not? That this exercise ought to be possible is further suggested by the fact that, for another feature of Conway's data, it can be done. After describing how different the environments of the separated twins often were, but how little this seemed to affect their IQ scores (and citing the case of the pair with IQs of 136 and 137), Conway states:

Only five have been brought up by relatives; and these, as it happens, include two pairs showing the largest discrepancies. For the whole group the correlation with the socio-economic ratings of the homes is admittedly positive, namely, 0.26; but the figure is only just significant. (Conway 1958, p. 186)

It is not immediately obvious what is being correlated with what here to yield a coefficient of 0.26. But some further clues provided in Conway (1959, pp. 7–8) make it clear that the correlation is between the difference in IQ scores within each of the 42 pairs of twins and the difference in the SES ratings of their homes. From the 53 pairs of twins shown in Table 3.6, it is indeed possible to select 42 which will yield a correlation of 0.26 between these two different scores.*

What, then, are we to make of all these discrepancies? We have already seen that, in his later years, Burt was prone to error. The table of correlations published in 1966 contains a number of unambiguous mistakes, and the most plausible account of this table, I have argued, suggests that Burt was quite capable of copying down numbers from the wrong places, so that no reliance can be placed on the numbers he reports. Should we accept that these new discrepancies provide evidence of yet further confusion, muddle and error, but nothing more sinister? This is possible, but there are grounds for scepticism.

1. The errors so far identified in Burt (1966) are confined to the table of correlations: indeed some of them are evident precisely because the numbers in the table are contradicted by numbers given in the text. Here, the discrepancies are with numbers given in the text of that paper (for example, the mean and standard deviation of the MZa twins' IQ scores).

2. There is an even larger discrepancy with the correlation for MZa twins reported by Conway (1958). Hitherto, we have seen no reason to suspect anything wrong with Conway's numbers. Her correlations for MZt and MZa twins all differ from those reported earlier in Burt (1955); her correlations for all other kinship categories remain unchanged, and it is obvious that the old numbers are simply being repeated. Everything seems in order. Must we now assume that the numbers in her table are just as prone to error as those in Burt (1966)?

3. There seem to be only two other possibilities. The first is that we have misidentified Conway's 42 pairs from the 53 pairs in Table 3.6. No doubt there is *some* combination of 42 pairs here that will yield the correct correlation. But this requires us to suppose that the explicit statements in Conway (1959) are simply false.

---

* In order to calculate this correlation, one has to know what to do with the cases marked R (residential institution) in Table 3.6. One solution is to assign them all a socio-economic status rating of 6. Another is to omit them from the analysis altogether. This does not make any difference: in either case, it remains possible to obtain a correlation of 0.26.

  It should be noted, however, that by no means all possible combinations of 42 twins from Table 3.6 yield Conway's correlation of 0.26. Most ways of achieving such a value assume that the pairs of twins *not* in her sample had relatively large discrepancies between their IQ scores. The effect of this restriction, of course, is to *increase* the value of the correlation between the IQ scores of the 42 twins assumed to be in her sample, so that the lowest possible value derivable from the numbers given in Table 3.6 is now slightly higher than the 0.901 I reported above.

4. The final alternative is to assume that the numbers sent by Burt to Jencks are in error. But there is a dilemma here. These numbers cannot be *seriously* in error, for in several respects they yield the right answer: a correlation coefficient virtually identical with that reported by Burt (1966); once plausible changes have been made to cases 1 and 53, the correct mean and virtually correct standard deviation for Burt (1966); and the appropriate correlation between differences in IQ and differences in SES for Conway (1958).

Taken together, I believe, these considerations suggest that Cronbach's smoking gun may well not be illusory. The reason why Burt took so long to send the data Jencks requested is that he did, indeed, have to calculate some or all of these data. He had published a correlation coefficient in 1966. Now he had to produce 53 pairs of scores that would yield this correlation coefficient. Some of these numbers may have existed. But some did not. They had to be calculated so as to produce the desired answer. At the same time, the numbers he produced had to satisfy other constraints: they had to be consistent with other things Burt (or Conway) had previously written about the twins. Even if Burt remembered everything he and Conway had written on the subject, it would simply not be possible (with a mechanical calculating machine!) to satisfy all these constraints at once in the time available. Hence some of the discrepancies.

Conclusion

Burt's MZ twin data and his other kinship correlations were among the first aspects of his work to attract serious criticism and veiled suggestions of fraud or other malfeasance. The charges concerning the invariant correlations and the missing assistants were the most widely publicized of all those brought against him, and probably remain the ones that have most seriously damaged his posthumous reputation. Is this fair?

It is clear that there is a great deal amiss with the 1966 paper in which Burt finally brought together all his data on twins and other kinship correlations. But it is not easy to decide on the true explanation of what is amiss. On some questions, I believe, there is little room for doubt. The main table of kinship correlations published in 1966 (Table 3.4 above) is riddled with error, and no faith can be put in any of these numbers. Nearly half the correlations are the same, to the third decimal place, as those reported earlier for different sized samples. There can be no totally innocent explanation of this that leaves Burt's reputation for meticulous integrity and careful science untouched. Coincidence can be dismissed out of hand. The defence that the majority of the invariant correlations appear for physical measurements, in which Burt was not particularly interested, and which he pointed out were not based on complete samples, does not withstand serious scrutiny. What Burt explicitly said about these correlations, and the samples on which they were based, is simply not consistent with the numbers he published. Either he was incredibly careless and muddled, or he was displaying a quite unacceptable disregard for the proper procedures for presenting scientific data.

There can be equally little doubt that neither Burt, nor any of his assistants, actually collected data on 32 new pairs of MZa twins between 1955 and 1966. Since this is the natural interpretation to put on the increase in sample size from 21 pairs in 1955 to 53 pairs in 1966, and since he talked of his collection of such twins having been enlarged and further cases having been brought to his notice, he is at best guilty of deliberately misleading his readers as to what had happened.

We have seen, however, good independent evidence that when he was preparing the tables for his 1966 paper, Burt was indeed capable of an astonishing degree of muddleheadedness. And it is at least conceivable that the 32 pairs of twins added after 1955 were not the result of collecting new data but of the rediscovery of old. One might argue, reasonably enough, that it seems odd that the old, mislaid data should have been rediscovered in such a piecemeal fashion; first 10 pairs, when the sample increased from 21 to over 30 pairs; then another 10 or so when it increased to 42 pairs; and a final 11 pairs between 1958 and 1966. But if it is conceded that the data were mislaid, it must also be conceded that they *might* have reappeared a little at a time, as first one box, then another, was sorted through.

Many of Burt's critics will regard this defence as implausible. They are surely entitled to do so. It is not, perhaps, a very likely tale, and Burt's clear deceptions serve only to increase one's suspicions. But even good grounds for suspicion fall short of definitive proof of fraud. If this is the strongest case that the prosecution can mount, then Burt probably ought not to be found guilty on this charge. He may be guilty, but the charge cannot be proved. I believe that the case for the prosecution is strengthened by Burt's reluctance to produce his raw data (the numbers sent to Jencks in 1969, shown above in Table 3.6, were the only set of test scores he ever sent to any of his correspondents, in spite of many requests for other data), by his reference to 'calculating' these data when all he needed to do was put them together, and by the demonstration that these numbers cannot be reconciled either with what Burt wrote in 1966 about all 53 pairs of twins, or with what Conway wrote in 1958 and 1959 about 42 of the pairs. Readers must decide for themselves how much weight to attach to these arguments. I think that the honest answer would probably have to be that they are sufficient to harden anyone's suspicions, but that they do not harden them into certainty. If the MZ twin data were the only grounds for accusing Burt of fraud, one would probably have to give him the benefit of the doubt.

References

Archer, G. (1983). Reflections on Sir Cyril Burt. *Journal of the Association of Educational Psychologists*, **6**, 53–5.

Banks, C. (1983). Professor Sir Cyril Burt: selected reminiscences. *Journal of the Association of Educational Psychologists*, **6**, 21–42.

Burt, C. L. (1943). Ability and income. *British Journal of Educational Psychology*, **13**, 83–98.

Burt, C. L. (1955). The evidence for the concept of intelligence. *British Journal of Educational Psychology*, **25**, 158–77.

Burt, C. L. (1958). The inheritance of mental ability; Bingham Lecture, 1957. *American Psychologist*, **13**, 1–15.

Burt, C. L. (1961). Intelligence and social mobility. *British Journal of Statistical Psychology*, **14**, 3–24.

Burt, C. L. (1966). The genetic determination of differences in intelligence: a study of monozygotic twins reared together and apart. *British Journal of Psychology*, **57**, 137–53.

Burt, C. L. (1969). Intelligence and heredity: some common misconceptions. *Irish Journal of Education*, **3**, 75–94.

Clarke, A. D. B. and Clarke, A. M. (1974). *Mental deficiency*, 3rd edn. Methuen, London.

Cohen, J. (1983). Sir Cyril Burt: a brief note. *Journal of the Association of Educational Psychologists*, **6**, 64–77.

Conway, J. (1958). The inheritance of intelligence and its social implications. *British Journal of Statistical Psychology*, **11**, 171–90.

Conway, J. (1959). Class differences in general intelligence. *British Journal of Statistical Psychology*, **12**, 5–14.

Cronbach, L. J. (1979). Review of Hearnshaw (1979). *Science*, **206**, 1392–4.

Fletcher, R. (1991). *Science, ideology, and the media: the Cyril Burt scandal*. Transaction Publishers, New Brunswick, NJ.

Fletcher, R. (1993). The "Miss Conway" story. *Bulletin of the British Psychological Society*, **6**, 214–15 (published posthumously).

Gillie, O. (1976). Crucial data was faked by eminent psychologist. *The Sunday Times*, 24 October. London.

Gillie, O. (1979). Burt's missing ladies. *Science*, **204**, 1035–8.

Hearnshaw, L. S. (1979). *Cyril Burt: psychologist*. Hodder & Stoughton, London.

Holzinger, K. J. (1938). Reply to special review of 'Twins'. *Psychological Bulletin*, **35**, 436–44.

Jensen, A. R. (1974). Kinship correlations reported by Sir Cyril Burt. *Behavior Genetics*, **4**, 1–28.

Joynson, R. B. (1989). *The Burt affair*. Routledge, London.

Kamin, L. J. (1974). *The science and politics of IQ*. Erlbaum, Potomac, MD.

McNemar, Q. (1938). Review of Newman, Freeman and Holzinger. *Psychological Bulletin*, **35**, 237–49.

Newman, H. H., Freeman, F. N., and Holzinger, K. J. (1937). *Twins: a study of heredity and environment*. Chicago University Press.

Sutherland, G. and Sharp, S. (1980). "The fust official psychologist in the wurrld": aspects of the professionalization of psychology in early twentieth-century Britain. *History of Science*, **18**, 181–208.

FOUR

Intelligence and social mobility
C. G. N. MASCIE-TAYLOR

Introduction

IN MAY 1961 the *British Journal of Statistical Psychology* published an article by Cyril Burt entitled 'Intelligence and social mobility'. Burt summarized his paper as follows:

The main thesis of the following paper is that, in a highly organized society, the discrepancies between the general intelligence of the children and the occupational class into which they are born is bound to produce a large and fairly constant amount of 'basic mobility', quite apart from any deliberate changes in the political or educational structure of the society.

Since the correlation between the intelligence of fathers and sons is only about 0.50 it is evident that, when classified according to their occupational status, (i) the mean intelligence of the children belonging to each class will exhibit a marked regression towards the general mean, and (ii) the intelligence of the individual children within each class will vary over a far wider range than that of their fathers. These deductions are fully confirmed by tables compiled to show the actual distribution of intelligence among adults and children belonging to the various occupational categories. It follows that, if the frequency distribution within the several classes is to remain constant (and still more if there is to be an increasing degree of vocational adjustment among later generations), a considerable amount of social mobility must inevitably take place, involving between 20 and 30 per cent of the population. Approximate estimates are attempted of both the actual and the ideal amounts. Data obtained from the after-histories of schoolchildren, followed up in later life, are analysed to ascertain the main psychological causes tending to produce a rise or drop in occupational status. (Burt 1961, p. 3)

Burt's paper led to increased research interest both in Britain and the USA on the relationship between IQ scores of fathers and their adult sons and occupational mobility. For example, in 1963 Young and Gibson published the results of a pilot study of 47 father–son pairs living in Cambridge, UK. Comparisons of the father–son IQs were in accord with Burt's findings, as were the results of a later study of Cambridge scientists, their brothers, and their fathers (Gibson 1970). In the USA support for Burt's findings was provided by Waller (1971). He used IQ data on 131 fathers and their 173 sons and found that the distribution of IQ scores by social class was relatively stable over the two generations and that differences between fathers and sons in social class were related to differences in their IQ scores.

The 1961 paper was quoted widely in both research papers and books. For instance, the eminent geneticists Professors L. L. Cavalli-Sforza and W. F. Bodmer cited the article in their text book *The genetics of human populations* published in 1971, as did the highly respected psychologist Professor I. Gottesman in his chapter 'Biogenetics of race and class' published in 1968. Eysenck (1971) and Herrnstein (1973) both relied heavily on Burt's paper for their analysis of IQ and social class.

A brief historical overview of the criticisms of the 1961 paper

By the mid-1970s many aspects of Burt's work were coming under close scrutiny and unfavourable reviews were beginning to appear. Kamin (1974, p. 155) criticized the 1961 paper on a number of grounds, including absence of information on how many adults and children were studied. McAskie and Clarke (1976) noted the absence of any explanation for the apparent decline in the mean IQ of the higher professional class from 153.2 as reported in Burt's 1943 article (Burt 1943) to 139.7 in the 1961 paper. In reviewing parent–offspring likeness for intelligence, McAskie and Clarke excluded a number of studies which they regarded as inadequate in one way or another. Referring to Burt they said 'The 1961 study is probably the most widely quoted statement on parent–offspring regression . . . However, careful inspection reveals a grossly inadequate description of the material and how it was collected. References to further details in other papers proved empty.'

Hearnshaw (1979) was also critical of the 1961 paper. He wrote 'Burt's reporting of his sources and methods was grossly inadequate' (p. 255), and 'What must be questioned are, firstly, the quality of the data themselves; and secondly, their use for fairly elaborate statistical analysis. It was a dubious exercise, though perhaps 'fraud' is too strong a word to use.'

Although Hearnshaw might have thought that 'fraud' was too strong a word, Dorfman did not. In the lead article in *Science* of 29 September 1978, Dorfman accused Burt of fabrication. The absence of any reference by Hearnshaw to Dorfman's paper is because the biography was in press by that time. Subsequently Hearnshaw did mention Dorfman's analysis when he gave a guest lecture delivered to the British Psychological Society in March 1980.

Dorfman's paper was entitled 'The Cyril Burt question: new findings' with a subtitle 'The eminent Briton is shown, beyond reasonable doubt, to have fabricated data on IQ and social class'. The summary to the article said, 'A detailed analysis of these data reveals, beyond reasonable doubt, that they were fabricated from a theoretical normal curve, from a genetic regression equation, and from figures published more than 30 years ago before Burt completed his surveys.'

Dorfman's allegations did not go unchallenged. The following February, Charlotte Banks (1979) in a letter to the *New Statesman* responding to an article on Burt by Oliver Gillie published the previous November, complained that Dorfman 'while quoting liberally from Burt's paper, omitted the crucial paragraph in the criticism'. Dorfman (1979a) stoutly defended his views in the same issue of the *New Statesman*. In April 1979 *Science* published two letters, one by Stigler and the other by Rubin. Stigler, in rebutting Dorfman's charges, wrote in his introductory paragraph 'Dorfman is in error on two major points, and his other points are sufficiently open to reasonable doubt to call his conclusion into serious question.' At the same time Stigler wished to distance himself from some of Burt's failings and his final paragraph says 'I do not wish to be interpreted as endorsing either Burt's statistical procedure or his unclear explanation of what he did (and his refusal to present the raw data). . . .' Rubin said much the same in his opening paragraph: 'Although I have no intention of defending Burt, I think that Dorfman's analyses provide no evidence for his claim. . . .' (Dorfman (1979b) replying in the same issue mainly addressed his remarks to the issues raised by Stigler and only briefly did he address Rubin's points. Later in the same year Rubin and Stigler (1979) wrote a joint letter to *Science* which criticized Dorfman's approach to data analysis and to which Dorfman (1979c) once more replied.

Since then others have offered their opinion on Dorfman's allegations. Hearnshaw sided with Dorfman in his 1980 lecture, saying 'There was also extremely doubtful material on parent–child correlations in intelligence, material which Dorfman's subsequent analysis (1978) has made look even more suspect.' Joynson (1989) and Fletcher (1991) however sided against Dorfman. Joynson said 'Neither position can be proved, but the Stigler–Rubin–Banks position is more likely. Burt's statistical techniques may well be open to objections, though these are not any clearer at the end than they were at the beginning.' Fletcher (1991) was more forthright when he wrote 'It is not true that Burt's statistics were proved faulty by Dorfman. Dorfman's own criticism was shown by Stigler, Cohen and others to rest on misunderstandings of a quite fundamental nature.' (The reference to Cohen refers to an apparently unpublished letter written by Arthur Cohen of New York to the editor of *Science* in 1978 which was also critical of Dorfman's allegations.) Thus the two most recent books on Burt suggest that Dorfman's evidence for fabrication was defective, a conclusion accepted by Jensen in his chapter in this volume (Chapter 1).

Details of the 1961 paper

Burt's 1961 paper was part of a wider debate on the causes of social class differences in intelligence. Burt held the view that a large part of the heterogeneity in intelligence between social classes was due to innate or inherited differences. In this respect he was at odds with some of the leading British sociologists (e.g. Floud and Halsey) who said that class differences in intelligence were the result of differences in environmental conditions (Halsey 1959). A large part of the introductory section of the paper covers points of agreement* and disagreement between Burt and his critics. But the empirical findings presented by Burt in the remainder of the paper were claimed to provide decisive evidence in support of the genetical (polygenic) theory.

These findings are the results of two quite separate sets of empirical analyses. In the first, Burt examines the relationship between social mobility and the intelligence of father–child pairs.** He shows the distribution of intelligence of adults in six occupational classes. There is a gradation in mean IQs from class I (higher professional) down to the unskilled classes (class VI) with an overall difference of nearly 55 points between the means of classes I and VI. This is followed by a table showing the distribution of intelligence for children based on the occupational classes of their fathers. Although there is still a gradation in means, the range for children is much smaller and the difference is only just over 28 IQ points between the means of classes I and VI.

Furthermore, the distribution of intelligence scores of children within each occupational class is wider than that of their fathers. Particular atten-

---

* It is noteworthy that in this section Burt draws attention (point 4) to the change in intelligence over time. He writes 'During the period for which information is available there has been no great change in the average level of general intelligence . . . On the whole, a survey of the relevant evidence would appear to suggest an actual but comparatively slight decline during the period in question, approximating to a drop of 1 or 2 I.Q. points per generation.' Assuming the 50-year period is equivalent to two generations, then a fall of up to four IQ points can be expected. Such a fall is not consistent with only a 'slight decline' or no great change. A decline in four IQ points is equivalent to a quarter of a standard deviation (if sd = 15) which implies a highly significant change in mean. It is strange that neither Burt nor the referees of the paper drew attention to this, more so given Burt's known statistical prowess and with the knowledge that he was publishing the article in a journal with strong leanings towards statistics. See Chapter 5 of this volume for a wider discussion of this issue.

** One assumes they were fathers and sons although they were mostly described by Burt as either parents or adults and children. Data on occupational class have traditionally (for fairly obvious reasons) been based on fathers rather than mothers; and Burt from time to time refers to sons or boys. For example 'the intelligence of fathers and sons' (1961, p. 3) and 'boys who belong to the highest occupational classes' (1961, p. 9).

tion is placed on this increased variability as evidence in support of the polygenic hypothesis. Burt wrote:

Consider, for example, the lowest occupational class of all. Among the adults only 20 persons out of 261 have an intelligence above the general average; among the children as many as 76, nearly four times as many—a discrepancy of 56. Dr Floud, and others who hold as she does that differences in intelligence are due wholly to environmental advantages or disadvantages, can hardly maintain the high level reached by these boys—all children of unskilled workers—results from the superior advantages which their home environments confer. But equally, those who adopt the traditional theory of blended inheritance, would find it quite impossible to explain the higher intelligence of these children in terms of heredity. On the Mendelian hypothesis, however, such apparent anomalies are exactly what we should anticipate if the amount of a child's intelligence is determined mainly, or at any rate largely, by his genetic constitution, and if that in turn is the result of a chance recombination of parental genes.

Burt also calculated the maximum amount of social mobility which could occur if occupational class depended solely on intelligence. After reducing the occupational classes from six to three groups by combining classes I to III, IV with V, and leaving VI separate, he demonstrated that just over 55 per cent of the adults are correctly placed, 23 per cent are in a group too high and should be moved down, while 22 per cent are in a group too low and should move up. He repeated the same exercise for the children and found only a third (33.5 per cent) of the children were correctly placed and approximately one third were placed too high and the other third too low. For the distribution of intelligence of the children to replicate that of the adults, 'steady state' social mobility must take place and Burt calculated this basic mobility as 22 per cent.

In the second set of analyses, Burt set out to see how intelligence, motivation, home background, and educational achievement influenced intergenerational occupational mobility. The analyses were based mostly on a different sample of children who had been followed up longitudinally and whose occupation was known (in Burt's words 'the after-school life'). Burt related mobility (simply categorized as up, none, or down) to four variables: intelligence (either sufficient to stay in that class, or above it or below it), motivation, home background and educational achievement (all either above or below the median). He concluded that intelligence and motivation were the most important causal factors in defining individual social mobility, although education and home background also made some contribution.

So both sets of analyses presented by Burt in the 1961 paper are supportive of the genetic theory; there is little to commend the environmental hypothesis. Such findings were of course very much in accord with Burt's own views.

The quality of the data

As I have already noted, a number of doubts have been raised about the quality of Burt's data, and the vagueness with which he described them. Questions were first raised 20 years ago by Kamin (1974) and soon thereafter by McAskie and Clarke (1976), Dorfman (1978), and Hearnshaw (1979). It is worth starting with some of the more minor areas of concern, leaving until later the more serious charges of fraud.

1. Lack of detailed description of the datasets

Burt refers to a number of datasets in the 1961 paper. In the introductory section he says the data 'are drawn from two overlapping inquiries or rather two series of inquiries'. The first inquiry refers to cross-sectional surveys of pupils in London schools while the second is of 'longitudinal studies of backward, gifted and normal pupils followed up into adult life'. As noted earlier, Burt describes the results of two separate analyses. The first set of analyses could well have been carried out on the cross-sectional data only, although this is never explicitly stated. The second set was carried out on the data originating from the longitudinal studies (Burt 1961, p. 19).

The surveys commenced in 1913 and they and the inquiries 'were carried out at intervals over a period of nearly 50 years'. Burt acknowledged that 'the data are too crude and limited for a detailed examination by a full analysis of variance' (1961, p. 9) but it is worth documenting just how imprecise he was in describing his sources of data.

For example, Burt writes 'For the children the bulk of the data was obtained from the surveys carried out from time to time in a London borough selected as typical of the whole county.' No evidence is provided by Burt to substantiate his claim of typicality. How likely is a London Borough to typify the whole county in the distribution of intelligence scores not only in 1913 but over a span of nearly 50 years? Given the changing social milieu of London boroughs it is very doubtful whether the distribution of intelligence in this unnamed London borough constantly mirrored that of the whole county over this period of time.

Information on the other datasets used in examining the extent of intergenerational mobility is also vague. Burt talks of 'fairly detailed data for just over 200 ordinary children who have reached an age when it is possible to say either that they have already moved out of their original class, in one direction or the other, or else that it is now practically certain that they will never do so. We have similar numbers for pupils who formerly attended central schools or won junior county scholarships as well as pupils who were educationally subnormal.' Burt then weights these subgroups to produce a 'composite group of males which shall be reasonably representative of the total population'. It is not clear how many children were followed up. One possibility is about 400, but the imprecise wording could imply that up to around 800 were studied.

Burt goes on to say:

For each of the sub-groups we have the following relevant information, obtained (except for vi) mainly when the children were at school: (i) the occupational class of the fathers at the time the children were born; (ii) assessments and descriptions of the home background, and particularly of the attitude of the family towards the child's social advancement; (iii) the child's own attitude, and particularly his industry, ambition, and educational and vocational aims; (iv) his level of intelligence, based on tests duly checked with the teachers and corrected where necessary; (v) his educational record (more especially his admission to a grammar school or its equivalent); (vi) his occupation when last visited. (1961, p. 19)

Although these six points are helpful, they leave many unanswered questions. The occupational distribution of fathers is not presented so there is no way of judging whether it is typical or otherwise. The exact number of children followed into adulthood is never stated. Is it reasonable to use the occupation of the father at the birth of the child as a correct measure of the class of origin? Some fathers might be expected to be mobile before attaining their main working life occupation. No details are provided on when and how home background, social advancement, a child's attitude, industry, educational and vocational aims were measured; whether they changed over time (presumably this would have been testable since this was a longitudinal study) and how attitude to social advancement was determined and whether it also changed over time. No information is given on what aspects of the educational record were used and how they were quantified. Were data from performance in school or national examinations available or was Burt relying on his own tests?

There is a paucity of facts about the occupation of the child in adulthood. How did Burt manage to keep track of these children into adulthood? At what age or ages were they followed up? Was the occupation at that time likely to be their main adult occupation? Why was Burt so confident it was? Where were they living—'last visited' might imply that they had remained in the locality and thus were easier to keep track of. If so, did this introduce any bias? No occupational distribution of the children is provided and so even after Burt has used fractional weights the reader has no way of judging whether the sample so obtained was representative or not.

2. *The description of how intelligence was measured in children and their fathers is poor*

Burt provided no specific information on which intelligence tests were used. When describing the second analyses he wrote 'the child's level of intelligence, based on tests duly checked with the teacher and corrected if necessary' (1961, p. 19). The corrections were made when children did better or worse than expected by the teacher so the scores presumably reflect 'adjusted' or 'final assessments' at best.

The way in which intelligence scores were obtained in the adults is also vague 'the data for the adults was obtained from parents of the children themselves' and 'for obvious reasons the assessments of adult intelligence were less thorough and less reliable' (1961, p. 9). Nowhere does Burt say exactly how intelligence was assessed in the fathers. In correspondence between Burt and Beardmore (quoted by Hearnshaw 1979, p. 255) Burt describes the slipshod way of assessing father's IQ.

3. Questions about the occupational classification used by Burt

In the first part of his studies, Burt presents data on the occupational classification of his sample of adults (presumably fathers). He wrote 'The occupational classification is much the same as that used in previous reports. It has been described by Carr-Saunders and Caradog Jones in their book on Social Structure in England and Wales.' Unlike the classification used in more recent studies of social mobility it is based, 'not on prestige or income, but rather on the degree of ability required for the work'. Burt goes on to give examples of the types of occupations in each of the six occupational classes, for instance 'class I includes those engaged in the highest type of professional and administrative work (university teachers, those of similar standing in law, medicine, education or the church, and the top people in commerce, industry, or the civil service).'

One might expect to find some explanation for the rationale for this classification in the book written by Carr-Saunders and Caradog Jones (1937). But this is not so. Carr-Saunders and Caradog Jones are merely citing Burt's own scheme. They wrote (commencing p. 55):

reference may be made to a most interesting calculation due to Professor Cyril Burt and revised later by him with the assistance of Miss Spielman. In fairness to Professor Burt it should be said that he regards his results as purely tentative. Occupations were classed into eight groups in descending order, according to intellectual attainments necessary if the duties involved were to be adequately performed. Thus for occupations placed in Group 1 it was judged that higher intellectual qualifications are required than those allocated to Group 2. As kindly described in a letter to us, the occupations given in the census were considered in turn and placed in its appropriate group after very careful inquiry as to the work involved. In a few cases it was found necessary to split up those grouped together in the census returns because some of them were doing work which required greater attainments than were demanded of others in the same group. The inquiry was limited to adult males, and when the allocation of occupations to their several groups was completed, it was easy to ascertain from the census the number of persons falling within each group and to calculate the percentage which the number in each group formed of the whole number of occupied adult males. . . .

Carr-Saunders and Caradog Jones then display a table of the eight occupational groups and the percentages in them. They conclude when referring to the table '. . . its author attaches no great weight to it.'

On the basis of this text it is hard to disagree with Dorfman's (1978) criticisms of Burt's classification scheme because of their lack of intrajudge or interjudge reliability. Moreover, there is good reason to question its accuracy. Burt's scheme is in fact relatively similar to one used by Glass (1954) in his analysis of 3497 British father–son pairs. But the percentages in the various occupational classes are quite different in Glass's and Burt's schemes. For instance, in Burt's data 26.1 per cent (261/1000) are in the unskilled group whereas Glass found only 11.1 per cent; for semi-skilled the percentages are 32.5 per cent (325/1000) and 13.1 per cent, and for the professional group 0.3 per cent and 3.7 per cent respectively. If one had used data reported by Carr-Saunders and Caradog Jones from the 1931 census, similar disparities would have been apparent. It is a mystery why Burt chose to continue to use a classification system in a paper published in 1961, which he had much earlier regarded as 'tentative', and to which he is reported to attach 'no great weight'. Further his scheme produces markedly different results from those of other classifications.

The fabrication issue

These points are sufficient to raise doubts about the adequacy of Burt's 1961 paper, and suggest that it should never have been treated with the respect it was. But they are not, of course, the main point at issue. That is simple: were Burt's data fabricated? I now turn to the various arguments raised under this heading. The claims and counterclaims as to whether Burt fabricated his data relate solely to the first set of analyses on the IQ scores of parent–child pairs. Neither Dorfman, nor his critics, discuss the inter-relationships between social mobility and intelligence, home background, motivation, and educational achievement analysed by Burt in the second, longitudinal part of his study.

1. The regression coefficients

The first contentious issue relates to the prediction of the mean IQs of sons in a class from the mean IQ of the fathers in the same class. Dorfman made use of a method cited by Conway (1959) to construct a formula for predicting the sons' mean IQ. Dorfman's formula was $\mu_{ic} = \alpha(\mu_{if} + 100)$, where μ_{ic} and μ_{if} are the population means of the children and fathers respectively for the ith occupational class and α is the regression coefficient.

Dorfman went on to calculate the regression coefficients for Burt's six occupational classes and found, after rounding to two decimal places, co-efficients of 0.50 in every class. Such results agreed perfectly with Conway's prediction from the genetic theory. Dorfman questioned whether such results should have been obtained from a 'pilot inquiry' where the data are 'crude'.

Stigler disagreed with the computations made by Dorfman and calculated regression coefficients based on a modification of a formula used by geneticists when determining the realized heritability* and said that one 'of Dorfman's major errors involves his calculation of regression coefficients . . . he has used the wrong formula. . . . Dorfman's formula is nonsensical'. Stigler could justifiably argue,** as calculations easily show, that all Conway was doing when calculating the offspring mean was to assume the parent–offspring regression was 0.5.†

However, what Dorfman (1979b) calculated, as he subsequently explained when replying to Stigler's criticisms, was not the statistical regression coefficient but what could be called a fabrication coefficient. Dorfman defended himself by pointing out he never defined the regression coefficient as a 'statistical regression coefficient' and to prevent confusion he did not use the normal sign for the regression coefficient β, but instead chose a different symbol α.

Although Dorfman's formula gives a coefficient of 0.50 when rounded to two decimal places, the actual coefficients to three decimal places were from class I to VI, 0.504, 0.497, 0.499, 0.502, 0.500, and 0.501. Thus using the rounded value of 0.50 would not yield the exact mean IQ of the sons in each class. However they are close, and Dorfman calculated an average absolute error of less than 0.4 of an IQ point. Furthermore the product–moment correlation between the six means was 0.999 which gave perfect support for the linear relation between the means as Dorfman had predicted.

One has to decide whether Dorfman's analyses are proof of fabrication. On the one hand the very close relationship between the mean IQs of fathers and sons (as measured by the correlation coefficient) and the nearly identical regression coefficients (computed for each occupational class) is, to say the least, highly suspicious. But in Burt's defence, one could reasonably argue that using the coefficient of 0.50 does not predict the mean IQ of sons except for those in class V, and there is therefore no case to answer.

---

* The formula is $R = h^2 \times S$, where S is the selection differential, h^2 is the heritability of the trait, and R is the response. Rewritten, the ratio $R/S = h^2$. (As Falconer points out, the term realized heritability denoted the ratio R/S irrespective of its validity as a measure of true heritability.)

** Stigler said, 'If we really wish to estimate the regression coefficients based on these limited data, we should presumably calculate $(X_c - 100)/(X_f - 100)$, which gives (to two decimal places) 0.52, 0.48, 0.49, 0.56, 0.50 and 0.49.'

† Dorfman quoted from Conway who wrote 'Thus, allowing for regression, and assuming that the I.Q. of the parents in the professional class averaged about 130, we should expect the I.Q. of their children to average 115. Similarly, if the intelligence of the "unskilled workers" averaged about 90, then we should expect that of their children to average about 95.' If we take Conway's first example assuming a population mean of 100, $S = (130 - 100) = 30$ and $R = 15(115 - 100)$ and $h^2 = 0.50$. Thus $S = (X_f - 100)$ and $R = (X_c - 100)$ which is the same formula as given by Stigler. Likewise for the second example $S = (90 - 100) = -10$ and $R = (95 - 100) = -5$. The $h^2 = -5/-10 = 0.5$. So in both cases $h^2 = 0.5$.

2. What was the sample size?

There has been much uncertainty as to how the figures which appeared in Burt's Tables I and II on the distribution of intelligence according to occupational class in adults and children were constructed. These tables, which are reproduced here as Tables 4.1 and 4.2, indicate the same total of 1000.

However, Burt never states the actual number of father–son pairs studied. Instead, in the last paragraph of section III, he explains what he has done:

In constructing the tables the frequencies inserted in the various rows and columns were proportional frequencies and in no way represent the number actually examined: from class I the number actually examined was nearer a hundred and twenty than three. To obtain the figures to be inserted (numbers per mille) we weighted the actual numbers so that the proportions in each class should be equal to the estimated proportions for the total population.

Dorfman (1979b) claimed the paragraph from which this quotation was taken was 'buried in a sea of deception and disingenuousness'. This is rather harsh criticism, and Burt makes a second reference on p. 16: 'Our method of reducing the figures observed to numbers per 1000 . . . .'. Even so readers of the summary would have had no idea that Burt has undertaken any transformation of the data since he wrote there of 'the actual distribution of intelligence among adults and children'.

The absence of any direct reference to sample size has led to a great deal of confusion. For instance Herrnstein (1973) originally believed that only 1000 father–child pairs were studied, but he later changed his mind and concluded that the number of pairs studied was 40 000. The latter figure, which is also quoted by Gottesman (1968) and Eysenck (1971), has clearly been calculated from the information given in the paragraph from Burt cited above, where he says the actual number in class I 'was nearer one hundred and twenty than three' (i.e. taking the upper figure of 120 gives a multiplier of 40 (120/3); hence $40 \times 1000 = 40\,000$). But there is no way of knowing whether the actual figure was 120, after all it could be any number equal to or greater than 62 (Kamin 1974). Moreover, we have absolutely no reason to believe that the weighting was the same for all cells; for some cells the weighting might have been smaller; in others Burt may have actually studied a smaller number of subjects than appear in Table 4.1.

Although Dorfman (1979b) was correct in saying that 'the consensus, however, has been that 40 000 father–child pairs were tested', that consensus is surely wrong as Kamin (1974) and Stigler (1979) have both pointed out. From what Burt says in his paper, there is simply no way of knowing the actual size of his sample. It might be a few hundred. It might be 40 000 (although one can share Kamin's scepticism at the thought of Burt actually testing 40 000 fathers). A possible answer was provided by

Table 4.1 Burt's adult data (From Burt 1961, Table I)

Distribution of intelligence according to occupational class: adults

		50–60	60–70	70–80	80–90	90–100	100–110	110–120	120–130	130–140	140+	Total	Mean IQ
I	Higher professional									2	1	3	139.7
II	Lower professional							2	13	15	1	31	130.6
III	Clerical				1	8	16	56	38	3		122	115.9
IV	Skilled			2	11	51	101	78	14	1		258	108.2
V	Semiskilled		5	15	31	135	120	17	2			325	97.8
VI	Unskilled	1	18	52	117	53	11	9				261	84.9
	Total	1	23	69	160	247	248	162	67	21	2	1000	100.0

Table 4.2 Burt's children data (From Burt 1961, Table II)

Distribution of intelligence according to occupational class: children

		50–60	60–70	70–80	80–90	90–100	100–110	110–120	120–130	130–140	140+	Total	Mean IQ
I	Higher professional						1		1	1		3	120.8
II	Lower professional				1	2	6	12	8	2		31	114.7
III	Clerical			3	8	21	31	35	18	6		122	107.8
IV	Skilled		1	12	33	53	70	59	22	7	1	258	104.6
V	Semiskilled	1	6	23	55	99	85	38	13	5		325	98.9
VI	Unskilled	1	15	32	62	75	54	16	6			261	92.6
	Total	2	22	70	159	250	247	160	68	21	1	1000	100.0

Burt in a letter (cited by Hearnshaw) written in reply to Beardmore who was seeking raw data on parent–child IQs:

1. For parents' intelligence the most reliable set of data I have consists of records of just over 100 fathers who attended L.C.C. schools as children between 1913 and 1920 and had their I.Q.s assessed. Their own children later attended L.C.C. schools and also had their I.Q.s assessed. . . .

2. I have a larger group in which each father's I.Q. was roughly assessed as an adult (about 370). The assessments were made by various social workers during their interviews; and the standards of the various interviewers were equated by getting them to apply camouflaged tests—oral questions ostensibly for information, and puzzles embodied in letters the fathers were asked to interpret. . . .

3. In making surveys from time to time of the schools in various boroughs, I again got attendance officers and social visitors to make rough estimates of fathers' I.Q.s and to obtain notes of their occupations. Pooling the whole lot together would yield well over 1000 cases. But the unreliability of the assessments would be very large. Since the errors probably range equally on either side of the true value, the figures are reliable enough for working out class averages.

This letter, if it provides a correct and truthful recollection by Burt of the datasets used in his 1961 paper, suggests a much smaller sample size. Even so, the letter does not explicitly state the number of pairs that could have been studied. The phrase 'pooling the whole lot' might refer to father–son pairs described under all three headings combined or, alternatively, to those under heading 3 only. Whichever of these interpretations is correct, the evidence of this letter suggests a modest sample size of at most 1500 father–son pairs.

The only other possible information is provided by an earlier paper (Burt 1955) where Burt reported a study of 1000 pairs of siblings and their parents, but stated that it had been possible to obtain IQ scores for only 954 parents. This study was the result of a series of surveys of London schools and it is hard to believe that the data reported there are not the same as those used for the 1961 paper.

Burt's letter to Beardmore not only sheds light on the sample size, it also shows how heterogeneous the datasets were, with some worthy of analysis, others less so. Hearnshaw is scathing in his denouncement:

This letter makes it quite clear that the material Burt dredged up from the past to serve as a basis for his calculations was, from a scientific point of view, mostly rubbish, and his admission that 'for obvious reasons the assessments of adult intelligence were less thorough and reliable' was a misleading understatement. Moreover he effectively concealed his sample sizes . . .

Subsequently Beardmore requested a copy of the data described under heading 1, but Burt did not forward any information. Hearnshaw believed that the failure to supply Beardmore with the data was because Burt no longer had the original information as it had been destroyed in the last

war. But Banks (quoted by Joynson 1989) felt that Burt's advancing years—
he was eighty-five years old at that time—might make him 'unwilling to go
to the attics to look for the data, especially as Ménière's disease made
bending down and looking for things very trying'.

3. Where do Burt's occupational class totals come from?

Tables 4.1 and 4.2 show the numbers of fathers and sons in each occupational class (Dorfman and others refer to these as the row totals). These are
not the actual numbers studied, but are weighted 'so that the proportions
in each class should be equal to the estimated proportions for the total
population' (Burt 1961, p. 10). Burt's description suggests that he had a
good idea of the proportions in the various occupational classes in this
London borough over the 40+ year period from 1913, thereby allowing
him to weight accordingly.

Whether Burt collected any such information is not known. But it is
extremely unlikely that he used such data when constructing the proportions in the different occupational classes. Instead, as astute detective work
by Dorfman (1978) has shown, Burt defined his totals using data from a
completely different source which bore little or no relationship to the
occupations of fathers in the 1961 paper.

In his Science paper Dorfman (1978) presented a table from a paper
published in 1926 by Spielman and Burt which shows the percentages in
eight 'vocational categories' (this table was subsequently reproduced by
Carr-Saunders and Caradog Jones with the words 'occupational group' re-
placing 'vocational categories'). The percentage for each occupational class
in the 1926 paper were remarkably similar to those which can be computed
from Burt's row totals in Table 4.1 as can be seen from Dorfman's Table 8
which is reproduced here as Table 4.3.

The main difference between the two sets of figures is for the unskilled
group in the 1961 dataset which was 26 per cent in the 1961 paper whereas
only 19 per cent were in that category in Spielman and Burt. However this
apparent discrepancy can be accounted for by summation of categories 6
and 7 (unskilled and casual labour). Dorfman rounded his percentages and
calculated the probability of getting complete agreement on all six percentages as 1 in 97 million. Dorfman concluded that 'Burt's 1961 row totals
were not the observed row totals per mille. Beyond any reasonable doubt,
they were taken by Burt from Spielman and Burt.'

As we have seen, Burt never claimed that his row totals were *observed*
row totals per mille. Burt wrote 'we weighted the actual numbers so that
the proportions in each class should be equal to the estimated proportions
for the total population' (1961, p. 10). Stigler and others accepted that the
estimated proportions were obtained from the Spielman and Burt article,
but defended Burt by saying that 'he weighted the counts to get precisely
the agreement that Dorfman presents as evidence of fabrication'. But, if

Table 4.3 Comparison of Spielman and Burt (1926) and Burt (1961) (From Dorfman 1978)

Percentages of male adults in the vocational categories given in Spielman and Burt (1926, Table III, p. 13). The percentages in levels 1 and 8 were rounded to whole numbers (the original figures are in parentheses). For comparison purposes, the percentage of adults in the six occupational classes in Burt's 1961 paper are similarly rounded.

Category level	Definitions		Percentage	
	Spielman and Burt[a] (vocational category)	Burt[b] (occupational class)	Spielman and Burt	Burt
1	Highest professional and administrative work	'Highest type of professional and administrative work'	(0.1) 0	0
2	Lower professional and technical work	'Lower professional or technical work'	3	3
3	Clerical and highly skilled work	'Intermediate types of clerical, commercial, or technical work'	12	12
4	Skilled work; minor commercial positions	'Skilled workers . . . commercial or industrial work of an equivalent level'	26	26
5	Semi-skilled work; poorest commercial positions	'Semi-skilled workers and those holding the poorest type of commercial position'	33	33
6	Unskilled labour and coarse manual work	'Unskilled labourers, casual labourers, and those employed on coarse manual work'	19 ⎫ 26	26
7	Casual labour		7 ⎭	–
8	Institutional cases (imbeciles and idiots)		(0.2) 0	–

[a] These definitions are precisely as given by Spielman and Burt (1926, p. 13) in their column 4 labelled 'Vocational category'.
[b] Taken from the text of Burt's 1961 paper (p. 10).

that were the case, surely one might have expected Burt to refer to the percentages as coming from Spielman and Burt or to Carr-Saunders and Caradog Jones in his text? No reference was made.

Furthermore Stigler and Burt's other defenders have missed an important point. The data for the 1926 article were based on (a) a mixture of married and unmarried adult males aged 14 years and older, some of whom had children, and (b) on Charles Booth's survey in the nineteenth century together with data collected from the 1921 census. Spielman and Burt state the percentages are 'nothing more than the roughest approximation'. Yet 35 years later Burt is quite content to apply these rough percentages, based on unmarried as well as married adults, with and without children, as estimators for a sample of fathers collected between 1913 and about 1960. Indeed Burt admits there have been changes in occupations over time. On p. 16 he writes 'The type of work available has changed appreciably . . .' and it is also likely that the qualifications and skills required for a particular occupation have changed.

Dorfman contends that the use of the Spielman and Burt percentages are proof of Burt's fabrication. In my view Dorfman is incorrect. One can question why Burt should use information gained from a quite different time period, with different sampling criteria, etc., as a basis for constructing the row totals in his 1961 paper. That Burt chose to use such data as estimators for the row totals is absurd, and many readers were not surprisingly misled by what he wrote. This is very bad science; but it is not proof of fabrication.

4. How normal were Burt's data?

A fourth charge of fabrication raised by Dorfman was that the intelligence scores presented in Tables 4.1 and 4.2 were remarkably close to a theoretical normal distribution. Dorfman made this claim having noted that Burt referred to the study as a 'pilot inquiry' and had acknowledged that the data were 'too crude and limited'. On the basis of these statements it might be expected that there would be only approximate fits of the father and son data to a normal distribution. On the contrary, Dorfman (1978) showed that the fit of Burt's data in Tables 4.1 and 4.2 to a theoretical normal distribution 'appears extraordinarily good' as Table 4.4 which is Dorfman's Table 4 shows.

Dorfman calculated the goodness of fit for both distributions and concluded that they did not differ significantly from the theoretical normal curve ($\chi^2 = 4.89$ for adults and $\chi^2 = 7.99$ for children). The untutored reader may find this unsurprising. But in fact, real data do not usually correspond so closely to the theoretical normal distribution. Dorfman went on to calculate the goodness of fit for a further 105 published distributions (of which 33 were for IQ, 42 for height, and 30 for weight distributions) and found that the data in Tables 4.1 and 4.2 were significantly closer to a

Table 4.4 Burt's rounded IQ distributions (column marginals) (From Dorfman 1978)

Burt's column marginals (Tables 4.1 and 4.2) as percentages rounded to whole numbers. For purposes of comparison, the theoretical normal distribution with mean 100 and standard deviation 15 [$N(100, 15)$] is also given.

	50–60	60–70	70–80	80–90	90–100	100–110	110–120	120–130	130–140	140+
Adults	0	2	7	16	25	25	16	7	2	0
Children	0	2	7	16	25	25	16	7	2	0
$N(100, 15)$	0	2	7	16	25	25	16	7	2	0

theoretical normal curve than the other 105 distributions. Dorfman concluded by saying

we may now say that, beyond a reasonable doubt, the frequency distributions of Burt's tables I and II were carefully constructed so as to give column marginals in agreement with the normal curve.

In large part, however, Dorfman's proof that Burt's data are too good to be true is based on a false assumption. In order to assess whether a distribution departs significantly from normality, or on the contrary is too normal, we need to know the sample size. Dorfman's calculations and comparisons with the other 105 distributions assumed that Burt's sample sizes were 40 000. If, on the other hand, we assume that the sample size was only 1000 then the data in Tables 4.1 and 4.2 no longer appear abnormally normal. This can be readily seen by visual inspection of Figure 4.1 in which the values of $\log_{10} (\chi^2/N)$ for fathers and sons with N each of 1000 have been added.*

Stigler (1979) raised other arguments as to why Burt's data showed such a close fit to a normal curve. He accepted that 'Dorfman demonstrates convincingly that Burt's column totals fit a normal distribution exactly, if rounding is allowed for', but he criticized Dorfman for not taking note of what Burt had written. On p. 10 Burt wrote 'Finally, for purposes of the present analysis we have rescaled our assessments of intelligence so that the mean of the whole group is 100 and the standard deviation 15.'

No doubt, as Banks also pointed out, this helps to explain why Burt's data are (approximately) normally distributed. But this is not the end of the problem. What did Burt mean by rescaling? The 1961 paper provides no clue. But in 1955, Burt stated the procedure he had followed: 'The actual measurements were transformed into standard scores (i.e., deviations divided by the standard deviation for each age); and these scores in turn converted to terms of an I.Q. scale with a standard deviation of 15.' In effect, Burt took his entire set of scores and transformed each individual IQ score so that the mean of the set was now 100 and the standard deviation 15. This is a perfectly legitimate procedure. But as Dorfman (1979b) argued, 'if Burt had rescaled the individual IQ's to fit a particular normal curve, he would have lost that normal curve by weighting the rows with proportions different from his actual sample proportions.'

* Dorfman chose to compute χ^2/N (which is asymptotically independent of sample size). He showed that in comparison with the 33 IQ distributions, the probability of obtaining a value as small as Burt's $\log_{10}(\chi^2/N)$ for the fathers and children was $p < 10^{-7}$ ($Z = 5.41$) and $p < 10^{-6}$ ($Z = 4.92$) respectively. He repeated the calculations for height and found probabilities of $p < 0.005$ and $p < 0.003$ for fathers and children respectively and for weight of $p < 10^{-20}$ for both fathers and children. With sample sizes of 1000 the Z values for IQ are 1.65 and 1.16 for fathers and children respectively which equate to p values of 0.0495 and 0.1230. (Dorfman does not state the mean and sd for χ^2/N but from reading the graph I calculated a mean of -1.60 and sd of 0.43 for IQ. These values are reasonably accurate since my computations with this mean and standard deviation gave Z values of 5.37 and 4.88 for fathers and children respectively which are close to Dorfman's published values of 5.41 and 4.92.)

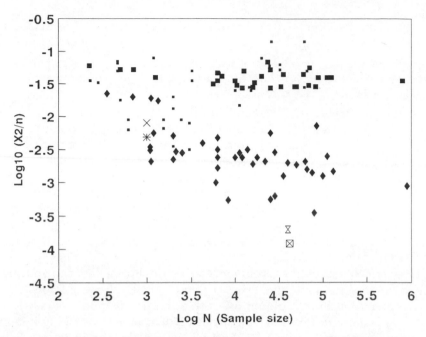

Figure 4.1 Log$_{10}\chi^2/N$ against log$_{10}N$ plotted by Dorfman (1978). Dorfman computed χ^2/N values for 33 frequency distribution of IQ, 42 for height, and 30 for weight. The χ^2/N values shown as ⊠ and ⊠ for Burt's data, children and adults, were calculated by Dorfman on the assumption of a sample size of 40 000. They are clearly different from the values of χ^2/N for all other distributions. The × and the ✳ show the value of χ^2/N for Burt's data on the assumption of a sample size of 1000.

It is easy to see why. Let us assume that Burt's total sample was actually 1000, but that 120 of this sample was in occupational class I. It follows that there must have been many fewer scores in some of the other classes than are shown in Tables 4.1 and 4.2. But if the original scores were transformed to a mean of 100, and standard deviation of 15, then as soon as Burt had reweighted his data, by reducing the proportion of those in occupational class I (with high IQs) and increasing the proportion of those in lower classes (with lower IQ scores), the mean of the reweighted sample would become less than 100 and the distribution skewed. Only if the actual proportions in Burt's sample were the same as Spielman and Burt's estimates would no changes occur (and that implies that Burt's sample was approximately 40 000 which we have good reason to doubt).

An alternative possibility, suggested by Stigler is that Burt rescaled his data by 'reweighting his columns (as he did his rows) to fit "estimated proportions" to a mean of 100 and standard deviation of 15'. But this also cannot be true. As Dorfman (1979*b*) noted, if Stigler was right then each cell mean would presumably have been at the midpoint of its IQ range and so the overall occupational class mean should equal the weighted sum of these midpoints. After excluding the three classes in Tables 4.1 and 4.2 where it is not possible to calculate the midpoint (because of 141+), only one of the nine comparisons was the same (children in class II).*

Dorfman (1978) also raised concerns about Burt's Tables III and IV which refer to the intelligence distributions of fathers and children which would occur if the occupational class depended solely on intelligence. Dorfman suggested that the very close similarity between the total numbers in each IQ range (the column marginals) for the children and fathers, and complete agreement if categories I and II are pooled 'seems extraordinary'. The probability of obtaining a value of χ^2 of zero is, of course, very small. Dorfman says 'Thus, beyond a reasonable doubt, Burt's assertion that "if we reclassify the actual data for adults according to these new borderlines, we obtain the distribution set out in Table III" is fraudulent. His χ^2's of zero are too good to be true.' Dorfman went on to show (a) that the data presented in Burt's Tables III and IV are not normally distributed, but (b) that it is possible to work out the figures in these tables from a normal distribution with mean of 100 and standard deviation of 15 by assuming that Burt used rounded Z values.

Dorfman's arguments seem plausible but again we have to contend with the vagueness of Burt's text. Dorfman places particular stress on what Burt said when introducing Tables III and IV: 'if we reclassify the actual data for adults . . . we obtain the distribution set out in table III', and 'Table IV shows the distribution of the children with the scale of intelligence sub-divided afresh'.

It is, indeed, not entirely clear how Burt obtained Tables III and IV from Tables I and II. The natural interpretation is that accepted by Dorfman: Burt took his distribution of 1000 rescaled IQ scores from each of Tables I and II and simply divided the scores at new boundaries. It is difficult to imagine what else could have been done. But there is one puzzling ambiguity. Burt's Tables III and IV are headed 'Rescaled', but Tables I and II are not. Can this mean that Burt rescaled his scores *again* when sub-dividing them afresh. This hardly makes sense, and is anyway inconsistent with Dorfman's observation that the scores in Tables III and IV are *not* normally distributed.

Even if we disregard the claims and counterclaims of Dorfman and Stigler, there is other evidence which suggests that Burt might have fabricated his data. This evidence was provided by Rubin (1979).** Rubin combined the

---

* For example, using class III (clerical) for fathers, the mean is calculated as 75×2 (midpoint of IQ range 70–80 multiplied by the number in that range)$+85 \times 11+95 \times 51+105 \times 101+115 \times 78+125 \times 14+135 \times 1=106.6$ which is different from the published value of 108.2.

** It is noteworthy that both Fletcher and Joynson have cited Rubin's arguments against Dorfman but neither mention Rubin's anti-Burt arguments.

information provided in Tables I to IV so as to calculate the IQ distribution in narrower categories, reproduced as Table 4.5.

Rubin demonstrated (a) that there were suspiciously small counts in the 90–1 category, much smaller than would have been predicted if the data conformed to a normal distribution (8 observed and 21 expected), and that (b) the distributions of fathers and sons were more similar to one another than either is to the normal distribution, and only one of the entries (children in the range 110–15) corresponded with the numbers predicted by a normal distribution. In particular, Rubin noted a bizarre feature of the fathers' and sons' IQ scores. If we ignore the 90–1 band, there is nearly always a difference between the numbers of fathers and of sons within each individual band. But when one combines neighbouring bands, e.g. IQs 50–60 with IQs 60–70, and IQs 70–80 with IQs 80–90, the number of fathers and sons becomes identical. In other words, the discrepancies in individual bands are virtually always precisely compensated for by an opposite discrepancy in the next band. It is difficult to see how this could have happened by chance. Rubin concluded:

if the IQ data are approximately N (100,15), the patterns in our Table 1 are suspicious, and if the IQ data are not approximately normal, the excellent fits of the IQ margins in Burt's tables I, II, III, and IV to the N (100, 15) model are suspicious.

He found an additional inconsistency when studying the narrower categories for each occupational group. Table I shows that in class VI there are 20 fathers (11+9) with IQs greater than 100, while in Table III Rubin calculated there were 24 with IQ scores more than 103. Rubin said, 'This inconsistency may be the result of trying to create the entries of tables from specified margins, as Dorfman suspects'; but he acknowledged that this inconsistency might be the result of a recording, computational, or typographical error, of which the latter is not unknown in Burt's publications.

Conclusions

Most people, though not all, who have written about Burt's work, and in particular his 1961 paper, agree that Burt provided very poor descriptions of his sources of data and techniques used. For example, Stigler wrote 'Burt's description of his procedure is extremely vague' and refers to 'his unclear explanation of what he did', while Rubin talked of 'Burt's ambiguous labelling of categories'. Dorfman took a stronger line: 'Burt . . . was extremely vague in his descriptions in order to mask his scientific fraud.' Whether or not Dorfman's strong inference is justified, there can be no reasonable doubt, as I have shown above, that Burt's description of his procedures etc. is wholly inadequate.

It is easy to say now that many of the ambiguities and uncertainties surrounding this paper could have been clarified had the editor of the

Table 4.5 Distribution of IQ scores for fathers and sons (From Burt 1961, Tables i–IV, as calculated by Rubin 1979)

IQ	50–60	60–70	70–80	80–90	90–91	91–100	100–103	103–110	110–115	115–120	120–127	127–130	130–140	140+
Fathers	1	23	69	160	8	259	86	162	96	66	56	11	21	2
Sons	2	22	70	159	8	242	83	164	94	66	56	12	21	1
Normal distribution	3	19	68	162	21	226	79	163	94	68	55	13	19	4

Combining neighbouring bands

IQ	50–70	70–90	90–103	103–115	115–127	127+
Fathers	24	229	333	258	122	34
Sons	24	229	333	258	122	34
Normal distribution	22	230	326	262	121	36

journal demanded publication of the raw data with clear and precise accounts of how modifications were carried out. It also demonstrates how easy it is for an editor, or someone very closely involved in the editorial process, to get an inadequate paper published. There is a case, as Thoday (1981) suggested, for editors to submit their own papers to external referees, nominated by someone other than the editor.

It is highly likely, as Dorfman pointed out, that the total numbers in each occupational class which appeared in Tables 4.1 and 4.2 were not based on the father's data but were simply weightings taken directly from Spielman and Burt's report. The absence of any reference to Spielman and Burt is castigated by Dorfman and dismissed by Stigler. Clearly Burt's use of Spielman and Burt is not an act of fraud, but it remains an absurd and wholly inappropriate thing to have done. Why should Burt have used these weightings since the basis for the data reported by Spielman and Burt bears little or no resemblance to those in the 1961 paper? Moreover the fact that Burt does not cite the source of his estimates in his 1961 paper suggests strongly that he knew that it was a wholly inadequate and inappropriate source.

The number of father–child pairs Burt studied remains a mystery. The majority of people who have written about the 1961 paper accept a figure of 40 000 father–child pairs. This number is clearly not true and is based on a misconception that a weighting of 40 should be applied to all cells. As Kamin noted 20 years ago, the weighting even for the one cell mentioned by Burt might be closer to 20 than to 40.

This uncertainty about the sample size is of some importance. Dorfman showed that the frequency distributions presented in Tables 4.1 and 4.2 fitted a normal curve remarkably well, and that they were very different from published frequency distributions for intelligence, height, and weight. The proof of the super-normality of Burt's distributions was based on the assumption of sample sizes of 40 000. However, use of more conservative sample sizes of 1000 reveals that the goodness of fit of the father and son distributions is in accord with the 33 IQ frequency distributions analysed by Dorfman. Thus the close fits to a normal curve exhibited by Tables 4.1 and 4.2 are not exceptional compared with other intelligence data when more modest and realistic sample sizes are used, and it would be wrong to use this as evidence of fabrication.

However there are other reasons to suspect that there is something wrong with Burt's data. Dorfman derived a formula which related the means of fathers and children. He used this formula to demonstrate that exactly the same relationship (to two decimal places) occurred in every occupational class. This is, to say the least, implausible. Furthermore the relationship between the father and child means was linear and the correlation was 0.999. Dorfman also revealed how easy it would have been for Burt to have fabricated the data and how easy it would have been for him to calculate the published means. Of course, the very high correlation is not proof of fabrication even if the data were supposedly crude and limited. Nor does the ease with which Dorfman fabricated the data prove that Burt did likewise.

The strongest evidence for fraud comes from two sources:

1. If Burt had normalized his data using the individual IQ scores of fathers and sons then his data would no longer have remained normal when the new weights from Spielman and Burt were used. Alternatively, if Burt normalized the data by cell means then the overall mean IQ of an occupational class would be equal to the weighted sum of the midpoints of each cell range. This is also not correct since only one out of the nine classes gives the correct mean. Thus neither of the straightforward interpretations of normalizing data can reproduce Burt's column totals.

2. Rubin was able to show, when combining the information from Burt's Tables I–IV, that the data were not as normal as hitherto supposed; there were too few counts in the range 90–1—only 8 when 21 would be expected. Furthermore Rubin's analyses showed that fathers' and sons' IQ distributions were more similar to one another than either was to the normal distribution. Even more suspiciously, he noted that digit discrepancies in one IQ band were regularly 'corrected' in the next. He also showed after reconstructing each occupational class using the narrower IQ categories 'a blatant inconsistency' in class VI. In Burt's Table I there are 20 with IQs greater than 100, while in Table III in the same class, 24 occur with IQs greater than 103. As Rubin acknowledges, this inconsistency might be evidence of Burt's fabrication but it could be due to some type of recoding, computational, or even, I suppose, typographical error.

So what can we conclude? There is no doubt, in my view, that Burt deliberately concealed information—on the sample size, on where the row totals came from, on when the information was collected. Even if he did not fabricate the data, then he was deliberately deceptive. But there really are good reasons to believe that the data were fabricated.

Science though, perhaps in a strange way, owes Burt a debt. His 1961 paper stimulated work on the study of IQs of father–son pairs and the relationship to social mobility. Much of what he believed—the inheritance of intelligence, the relationship between social mobility and IQ has been confirmed by studies in a number of countries. The removal of Burt's suspect data still leaves a large body of data showing mean differences between social classes and other groups. Such differences, which exist for a variety of characteristics, not just IQ, are due to the action and interaction of genes and the environment.

References

Banks, C. (1979). Correspondence. *New Statesman*, 2 February, **150**.

Burt, C. (1943). Ability and income. *British Journal of Educational Psychology*, **13**, 83–98.

Burt, C. (1955). The evidence for the concept of intelligence. *British Journal of Educational Psychology*, **25**, 158–77.

Burt, C. (1961). Intelligence and social mobility. *British Jouranl of Statistical Psychology*, **14**, 3–24.

Carr-Saunders, A. M. and Caradog Jones, D. (1937). *Social structure in England and Wales*. Clarendon Press, Oxford.

Cavalli-Sforza, L. L. and Bodmer, W. F. (1971). *The genetics of human populations*. Freeman, San Francisco, CA.

Conway, J. (1959). Class differences in general intelligence. *British Journal of Statistical Psychology*, **12**, 5–14.

Dorfman, D. D. (1978). The Cyril Burt question: new findings. *Science*, **201**, 1177–86.

Dorfman, D. D. (1979*a*). Correspondence. *New Statesman*, 2 February, **150**.

Dorfman, D. D. (1979*b*). Reply to Stigler and Rubin. *Science*, **204**, 246–54.

Dorfman, D. D. (1979*c*). Burt's data: Dorfman's analysis. *Science*, **206**, 142–44.

Eysenck, H. J. (1971). *The I.Q. argument: race, intelligence and education*. Library Press, New York.

Fletcher, R. (1991). *Science, ideology and the media: the Cyril Burt scandal*. Transaction Publishers, New Brunswick, NJ.

Gibson, J. B. (1970). Biological aspects of a high socio-economic group. I. IQ, education and social mobility. *Journal of Biosocial Science*, **2**, 1–16.

Glass, D. V. (1954). *Social mobility in Britain*. Routledge and Kegan Paul, London.

Gottesman, I. I. (1968). Biogenetics of race and class. In *Social class, race, and psychological development* (ed. M. Deutsch, I. Katz, and A. R. Jensen), pp. 11–51. Holt, Rinehart and Winston, New York.

Halsey, A. H. (1959). Class differences in general intelligence. *British Journal of Statistical Psychology*, **12**, 1–4.

Hearnshaw, L. S. (1979). *Cyril Burt, psychologist*. Hodder and Stoughton, London.

Herrnstein, R. J. (1973). *I.Q. in the meritocracy*. Little Brown, Boston, MA.

Joynson, R. B. (1989). *The Burt affair*. Routledge, London.

Kamin, L. (1974). *The science and politics of IQ*. Erlbaum, Potomac, MD.

McAskie, M. and Clarke, A. M. (1976). Parent–offspring resemblances in intelligence: theories and evidence. *British Journal of Psychology*, **67**, 243–73.

Rubin, D. B. (1979). Reply to Dorfman. *Science*, **204**, 245–6.

Rubin, D. B. and Stigler, S.J. (1979). Dorfman's data analysis. *Science*, **205**, 1204–6.

Spielman, W. and Burt, C. (1926). In *A Study in vocational guidance*, Report No. 33 (ed. F. Gaw, L. Ramsey, M. Smith, and W. Spielman), pp. 12–17. HMSO, London.

Stigler, S. J. (1979). Reply to Dorfman. *Science*, **204**, 242–6.

Thoday, J. M. (1981). Probity in science: the case of Cyril Burt, review of Beloff. *Nature*, **291**, 517–18.

Waller, J. H. (1971). Achievement and social mobility: relationships among IQ score, education, and occupation in two generations. *Social Biology*, **18**, 252–9.

Young, M. and Gibson, J. B. (1963). In search of an explanation of social mobility. *British Journal of Statistical Psychology*, **16**, 27–36.

FIVE

Declining educational standards
N. J. MACKINTOSH

Burt's views on education, like most of his views, remained remarkably consistent throughout his life. Well in advance of their time when he entered the educational world before the First World War, his views did not alter with developments in educational theory and practice, and by the end of his life they were out of tune with many so-called 'progressive' trends of the time. (Hearnshaw 1979, p. 122)

HEARNSHAW'S JUDGEMENT is perceptive and fair. Burt was, for example, a firm believer in equality of educational opportunity, and consistently argued that the English educational system was unfairly biased in favour of children from middle-class backgrounds who gained access to grammar schools and universities far more readily than their innate abilities justified. As he frequently pointed out (e.g. Burt 1939, 1943, 1969a), the average IQ of middle-class children might be, indeed was, higher than that of working-class children, but the fact that there were so many more of the latter than of the former meant that there were actually more working-class than middle-class children with an IQ of, say, 115 or more. In spite of which, the vast majority of university students came from middle-class backgrounds.

If this might, even today, be regarded as a reasonably liberal point of view, some of Burt's other attitudes towards educational selection, undoubtedly progressive in the 1920s and 1930s, were widely denounced as reactionary in the 1950s and 1960s by progressive educational thinkers and politicians. Whatever Gould (1981) or Kamin (1981) may say, Burt was certainly not responsible for the institution of the 11+ exam in English schools after the 1944 Education Act, let alone for the practice of selection for 'free places' in secondary schools in the 1920s and 1930s. Selection was already built into the system: the only question at issue was the basis on which it was to occur (Sutherland 1984). Educational psychologists, such as Godfrey Thomson and Burt, argued that IQ tests were likely to produce a fairer system of selection than parental interview, teachers' assessments, or even tests of written English—all of which favoured children from middle-class backgrounds. But by the 1950s, selection for secondary education in England came under increasing attack, and Burt, who continued vigorously to defend it (e.g. Burt 1959), was soon cast in the role of reactionary bogeyman, championing an outmoded, unjust, and elitist practice. There seems little doubt that Burt enjoyed this sort of controversy, and this perhaps helps to explain why he also, at about this time, started attacking other educational trends, notably the progressive methods of education in primary schools, which were endorsed by the Plowden Committee (1967), but which Burt believed to be based on an uncritical acceptance of the work of Jean Piaget, and to have resulted in a decline in certain educational standards.

In 1969, in a contribution to the second Black Paper on education, Burt wrote:

In the Plowden Report we are told that 'the gloomy forebodings of the decline in knowledge that would follow progressive methods have been discredited.' However, educationists today no longer seem quite so sure as some of them were when the report was compiled . . . Certainly, as the Plowden Committee was able to show, there has been during the last 25 years, a marked improvement in the level of attainment in the basic subjects. But, owing to the evacuation and the call-up of teachers, the level reached at the end of the last war was at its lowest ebb. If we go back to the period just before that war, or again just before the first world war, the overall trend has shown, not an improvement, but, if anything, a decline. (Burt 1969a, p. 23)

Burt documented these claims in an article published in the *Irish Journal of Education* (Burt 1969b). In fact, he introduced his discussion of his evidence here in a reasonable and temperate manner. He acknowledged that progressive policies had resulted in

a pronounced improvement in educational methods in the infants school, particularly in regard to the rudiments of reading and number. In some of the primary schools remarkable results have also been achieved by a few devoted and enterprising teachers . . . The classroom is no longer a mental treadmill; it has become a happy place. And this is no small gain provided work and self-discipline do not suffer. (p. 87)

Nevertheless, [he continued] when we turn to the average accomplishments of the school population as a whole, we search in vain for evidence of any marked improvement either in the basic educational subjects or in general ability. (loc. cit.)

As far as general ability was concerned, he then stated that

In spite of the vast improvement made in social conditions during the last fifty years and the alleged improvements in educational methods, there are no signs whatever that the average level of intelligence has been raised. Nor has there been any discernible levelling up of the intelligence of the duller children. The mean IQ has remained at about 100 and the standard deviation at about 15 or 16 on the revised Binet scale, with minor fluctuations well within the margin allowed by the standard error. (op. cit., p. 88)

Table 5.1 Comparison of school attainments, 1914–65 (from Burt 1969*b*)

Year	Intelligence	Reading		Spelling	Arithmetic	
		Accuracy	Comprehension		Mechanical	Problems
1914	100.3	101.4	100.1	102.8	103.2	101.3
1917	100.1	95.3	96.5	94.7	91.1	92.5
1920	100.0	100.0	100.0	100.0	100.0	100.0
1930	98.6	100.7	105.2	100.8	103.4	94.7
1945	99.3	90.8	91.1	89.5	88.9	93.2
1955	99.5	96.7	99.4	94.6	95.5	97.6

The data on which this claim was based are shown under Intelligence in Table 5.1 above. I shall return to them shortly, after discussing his evidence on school attainments. Here he wrote:

For a comparison of school attainments I am indebted to a study carried out by Miss MG O'Connor. She has compiled data from various surveys and reports from 1914 onwards, based on tests applied by teachers or research students. They relate to the last year of primary school (age 10 to 11). The data are presented in Table 1 [and are reproduced as Table 5.1 here]. The figures in the table are medians; those obtained in 1920 (the year of the survey reported in *Mental and Scholastic Tests* [Burt 1921]) are taken as 100. The most striking feature that emerges is the zig-zag fluctuation in each of the subjects tested, never very large, and due mainly, it would seem, to the effects of the wars and the subsequent recovery in each case. As the Plowden Report and other investigations have amply demonstrated, there has been, since the end of the last war, a substantial improvement in the basic subjects—most of all in comprehension of reading. Yet even so, especially where accuracy is concerned, the level reached in each of the three R's is still below that which was attained in 1914, when teachers concentrated almost all their effort on these fundamental processes. If we took the medians for that year as standard, then the decline would be still more obvious: the figures for spelling would be only 91.1 and for mechanical arithmetic 92.5. (op. cit., p. 89)

Hearnshaw's doubts

Miss M. G. O'Connor has proved to be an even more mysterious figure than Miss Howard or Miss Conway. As Hearnshaw notes,

there is no evidence of any contact or communication with Miss MG O'Connor. She never appears to have visited Elsworthy Road [Burt's flat], nor to have written to Burt, nor to have met any of Burt's associates. (Hearnshaw 1979, p. 259)

Fletcher acknowledges that

Miss MG O'Connor does remain a third missing lady. She has not yet been traced. While there is substantial evidence of various kinds for the actual existence of Miss

Howard and Miss Conway, there seems to be none so far for the existence of Miss O'Connor. (Fletcher 1991, pp. 293–4)

That remains true today: the nearest anyone has come to tracing her is Richard Rawles's discovery of a Miss N. M. O'Connor, who was a member of the British Psychological Society from 1924–36 (Rawles 1977). Apart from the difference in initials, it does not seem particularly plausible to suppose that, aged well over 60, this is the person who was compiling data for Burt on school attainments in the late 1960s. But it is not this that led Hearnshaw to suspect that at least some of the numbers that appear in Table 5.1 were simply invented. His argument was rather different. How, he asked, had these results been collected?

The years 1917, 1920, 1930 and 1945 present no real problems, as Burt was working in London then, and had access to schools . . . The figures for the remaining years, 1914, 1955 and 1965, are not so easy to explain. Burt no doubt had tested a lot of children in 1914 . . . but at that early stage of his career he had no standardised tests of scholastic attainments. . . . The position with regard to the years 1955 and 1965 is even more problematic, because by then Burt had no right of entry to London schools . . . So it seems improbable in the extreme that any testing programme was carried out under Burt's direction either in 1955 or 1965 . . . The whole massive operation involved in ascertaining changes in standards of scholastic achievement left not a trace in Burt's detailed diary entries, nor in his carefully filed correspondence. The conclusion seems inescapable: the figures given in Burt's table in the *Irish Journal of Education* were, at least in part, fabricated. (Hearnshaw 1979, pp. 257–9)

Both Joynson (1989) and Fletcher (1991) dismiss Hearnshaw's charges, the latter regarding the whole matter as 'something of a tempest in a teacup' (Fletcher 1991, p. 289). According to Joynson, the 1914 results may indeed have been based on inadequately standardized tests given to small pilot groups, but that

does not amount to fabrication of data, only to a use of unreliable data—and in view of Burt's warnings about the 'precarious' nature of his comparisons, this provides no evidence of dishonesty. (Joynson 1989, p. 208)

As for the 1955 and 1965 data, Joynson argues, no massive operation was involved—and none claimed. These (and perhaps other data in the table) were obtained by a method of 'median sampling', which Burt described in his paper as follows:

The school inspectors can usually be relied on to select average or median schools; and in each of these the investigator then picks out the median pupil in the age-group under consideration . . . Actual trial shows that by testing about 20 or 30 children thus selected (often even less) one can get a much better estimate of the average of the general population than if one attempted to test all the children in what was designed to be a genuinely random sample. (Burt 1969*b*, pp. 88–9)

It is quite clear from the context here that Burt was describing this procedure as a solution to the problem of getting accurate measures of *intelligence*, where 'individual tests' such as the revised Binet and standardised performance tests, must be used' (loc. cit.), and even here he was not claiming that any of the data appearing in Table 5.1 were actually collected in this way: the passage appears after Burt has said how desirable it would be if similar surveys could be carried out elsewhere and that this need not be as difficult as some people might suppose. Nevertheless, it is entirely possible that some of the 'various surveys and reports' of educational attainments, 'based on tests applied by teachers or research students' were indeed obtained by this method of median sampling—which Burt had originally described in *Mental and scholastic tests* (Burt 1921). A characteristic feature of Table 5.1, readily acknowledged by Joynson, is that Burt provides no details at all of the samples for *any* year and, as we shall see, only the sketchiest details of the actual tests employed on any given occasion.

An alternative possibility, suggested by Audley and Rawles (1990), is that the 1955 and 1965 data were provided by the National Foundation for Educational Research (NFER). Although Hearnshaw was sceptical of any such possibility, since 'Mr DA Pidgeon of NFER was one of Burt's principal critics on the matter of educational standards' (Hearnshaw 1979, p. 258), according to Dr Clare Burstall, who was Director of NFER until 1993, Burt certainly would have had access to NFER survey data after the war:

Pidgeon used regularly to discuss the national survey data with Sir Cyril who was, at that time, a Vice-President of NFER and thus entitled to receive the data on children's educational attainment. Douglas Pidgeon . . . used frequently to seek Sir Cyril's advice on the treatment of the national survey data. (Personal communication from Dr Clare Burstall, 16 Feb, 1992.)

The real problem

On the question whether Burt *could* have obtained the data shown in Table 5.1, we should perhaps accord him the benefit of the doubt. The data could have come from a variety of different sources, and may not have represented the 'massive operation' Hearnshaw supposed. One might even argue that Burt's protestations about the 'precarious' nature of his comparisons, and his acknowledgement of various 'sources of inaccuracy' (1969*b*, p. 90), suggest precisely that he knew full well that the data were quite inadequate to justify any claims for the reliable and representative nature of his samples. But this does not end the matter. Indeed, I believe that Hearnshaw's suggestion has distracted attention from the central reason for doubting the validity of the data shown in Table 5.1. The critical question is not whether Burt could have collected, or had access to, such data. It is whether any meaningful survey or surveys could possibly have produced the pattern of results shown in Table 5.1. And that question arises in its most acute form, not for the results on school attainments, but for those on intelligence.

Why should this be? The problem is this. Let us suppose that we were possessed of reliable test results for, say, arithmetic problem solving, for representative samples of 10-year-old London schoolchildren, for the years 1914 to 1965. Would such results be sufficient to establish whether educational standards had risen or fallen over this period? The answer must in fact be: no. In order to know whether children's attainments have or have not changed, we should also need to be assured that comparable and appropriate tests had been used in each year. A moment's reflection suggests that there is something of a dilemma here. We could, of course, ensure that the test used in 1965 was *comparable* with the test used in 1914, by using exactly the same test on both occasions. But would it be *appropriate* to give a 1914 test in 1965? The finding that 1965 10-year-olds did worse on this test might mean only that they had been taught arithmetic in rather different ways. Perhaps the 1914 10-year-olds would have obtained worse scores than those born in 1955 if both groups had been tested on a 1965 test.

A similar dilemma arises if we wish to ascertain whether there have been any changes in intelligence over the 50-year period from 1914 to 1965. Do we ensure that the tests used are all comparable—for example, by using exactly the same test on each occasion? Or do we ensure that they are all appropriate, by using a newly standardized test on each occasion? What did Burt do? He certainly did not follow the second course, since this would have been a quite pointless exercise. The average IQ of the population is 100, which is to say that if a truly representative sample of the population is given a recently standardized IQ test, their mean score will be 100. If it is not, then either the sample was not representative, or the test was not properly standardized. If this is what Burt was reporting in the column headed Intelligence in Table 5.1, then all the scores in that column should be approximately 100 (as on the whole they are), but this would tell us precisely nothing about possible changes in intelligence over this 50-year period. The only point in Burt's presenting such data would be to assure us that his samples were representative. But this is *not* what he says:

In spite of the vast improvement made in social conditions during the last 50 years and the alleged improvements in educational methods, there are no signs whatever that the average level of intelligence has been raised. (Burt 1969*b*, p. 88, see above for the continuation of this passage)

Burt is explicitly *comparing* levels of intelligence over this 50-year period, and thus implicitly claiming that he used comparable tests on each occasion. There are, in fact, two rather different ways in which he might have done this. He could simply have used exactly the same test or set of tests on each occasion. Alternatively, he could have used different tests on each occasion, but ensured that they were comparable by translating the norms of one into those of another. This requires something like the following procedure. We know that a representative sample of 10-year-olds born in 1955 will obtain an average score of 100 on a test standardized in 1965, just as a similar sample born in 1920 will obtain a similar score on a test standardized in 1930. We can find out whether the tests are comparable by giving the children born in 1955 the 1930 test. If their average score is 100, the two tests have similar norms and that is the end of the matter. But if it is 110, the 1930 test is 10 points easier than the 1965 test, and the latter's norms must be adjusted by 10 points to make them comparable with the earlier test.

It should be clear, therefore, that assessing changes in intelligence over a 50-year period is not an entirely straightforward matter. For the exercise to be meaningful, it requires some specification of the tests used; more particularly, we need to know whether the same test has been used on every occasion, and if not, what comparisons have been used to allow the translation of the norms of the later tests into those of the earlier tests. It will come as no surprise to learn that Burt is not quite as specific as one might wish on these points. Here is what he says:

On my appointment in 1913 as Psychologist in the London schools, I commenced, with the aid of the teachers, regular surveys by means of standardised tests both of intelligence and of school attainments. They were repeated at intervals of three years and later on at intervals of about ten years. The tests and sampling methods we adopted were described in *Mental and Scholastic Tests* [Burt 1921]. Since the war similar studies have been carried out by my own research students and by various other investigators. Many of them have used the same standardised tests: others have preferred those of Schonell, Vernon, Wechsler, or the National Foundation for Educational Research [some of these are tests of school attainment rather than of intelligence]. In these latter cases the results recorded can readily be translated into terms of the earlier norms. When checking the standards for the latest edition of my book [*Mental and scholastic tests*] in 1962, I collected a good deal of fresh data, and was surprised to find that in many respects the changes seemed comparatively small. (Burt 1969b, pp. 87–8)

Although this does not banish all uncertainty, the implication seems reasonably clear. The data in Table 5.1 come from a mixture of both procedures. Until 1930 at least, the same tests, those detailed in *Mental and scholastic tests*, were used on each occasion. Some of the intelligence data for 1955 and 1965 (and perhaps those for 1945) may also come from these same tests; some may come from more recent tests, but where this was so, their norms were translated on the basis of the 'good deal of fresh data' collected for the 1962 edition of *Mental and scholastic tests*. In fact, Burt was not exaggerating when he said that the changes needed for this edition were comparatively small. He made none at all: the norms for the 1962 edition (as for all earlier editions) were exactly the same as those of the original 1921 edition.

We are now, however, in a position to pose the critical problem. If Burt had actually obtained his data on intelligence in this way, then it is virtually impossible that he would have found 'no signs whatever that the average level of intelligence has been raised'. On the contrary, he should have found clear signs of a substantial increase in the average level of intelligence—for that is precisely what has happened. Thanks to the meticulous research of James Flynn (1984, 1987) ably followed up by Richard Lynn (Lynn and Hampson 1986; Lynn *et al.* 1987; Lynn *et al.* 1988) there is now incontrovertible evidence that, in virtually every industrialized country in the world, including Britain, there have been striking improvements in performance on standardized IQ tests over the past 60 years or more. In his initial study, Flynn (1984) calculated that performance had been improving in the USA at a rate of about three IQ points per decade between 1930 and 1980; his subsequent study (Flynn 1987) suggested that the rates of increase had been even greater in some West European countries since the war, although rather less in Britain.

Flynn relied on only three studies for his conclusions concerning Britain, one of which (the restandardization of Raven's matrices) he rightly argued probably did not provide comparable samples on each occasion. The other two studies, however, are excellent (Cattell 1950; Scottish Council for Research in Education 1949). In both cases, large representative samples of 10–12-year old schoolchildren (in Cattell's study the population of Leicester, in the SCRE study the population of Scotland) were administered the same set of tests, first during the 1930s, secondly shortly after the Second World War. The later generation performed slightly but significantly better than the earlier: across both studies the rate of improvement was 1–2 IQ points per decade. But as Lynn and Hampson (1986) had noted, Burt might have been right in supposing that the war had a depressing effect on children's test performance: if that is so, these studies would have underestimated the improvement occurring at other times. Indeed, a study by Pilliner *et al.* (1960) suggested that there had been an improvement of six IQ points between 1945 and 1957 in 11-year-olds' performance on verbal reasoning tests, and Lynn *et al.* (1987) confirmed that the rate of increase in England since the war has probably averaged at least three points per decade.

Although these studies cited by Flynn and Lynn provide data on test scores only since 1930, there is one American study which suggests that increases in performance go back even further. Tuddenham (1948) compared US military test scores from the First World War with those obtained in the Second World War, and calculated that the later generation outscored the earlier by over ten IQ points. A gain of this magnitude could not all have occurred during the 1930s, and the remarkable implication is that test scores have been improving ever since IQ tests were first invented. Even a conservative assessment of the average rate of increase in Britain at two points per decade implies that Burt should have observed a ten-point gain between 1914 and 1965, rather than a one-point decline. Moreover, it seems probable that much of that gain should have occurred in spurts after each war, with relatively little gain over the war years themselves. There is absolutely no evidence of such a pattern in Burt's data, which show a virtually uniform set of scores, with a modest dip in 1930. As I argued above, therefore, the question at issue is not whether Burt was in a position to collect the data published in his 1969 paper; it is why there is such a discrepancy between Burt's data and what we have every reason to believe is the true pattern of results.

Possible explanations

What is the explanation of this discrepancy? Several possibilities can be suggested, some entirely benign, others, which I shall leave until last, progressively less so.

1. Increases in test scores have not been uniform for all IQ tests. Perhaps the particular tests that Burt used yielded smaller gains than others would have. Flynn (1987) found that the largest gains had occurred in non-verbal tests such as Raven's matrices, and that within the Wechsler tests (WAIS and WISC) gains were more marked on the non-verbal performance scales than on the verbal scales. There is some evidence for a similar discrepancy in Britain. Between 1932 and 1947 in Scotland, the rate of increase on a verbal test was 1.5 points per decade (SCRE 1949), but Lynn *et al.* (1987) reported an increase of 2.5 points per decade over approximately the same period on the Cattell non-verbal tests. Set against this, the six-point increase between 1945 and 1957 reported by Pilliner *et al.* (1960) was on a verbal reasoning test.

It is, of course, unfortunate that Burt did not specify unambiguously all the tests he relied on, especially after 1945. But the earlier tests were, presumably, 'the revised Binet and standardized performance tests' from *Mental and scholastic tests* (Burt 1921). Although the Binet tests are predominantly verbal, they also contain non-verbal items (e.g. drawing shapes from memory) and Flynn's analyses suggest that in the US, gains on the Stanford–Binet test have been comparable with those on the WAIS and WISC as a whole, rather than with those on their verbal scales alone. There is, of course, no reason to suppose that Burt's performance tests would have yielded smaller than average IQ gains.

2. Perhaps British increases in test scores were more pronounced in some parts of the country than in others. Burt's data presumably all come from London, and it is conceivable that changes in the population of London caused the IQ of London schoolchildren to fall behind as other parts of the country forged ahead. The population of London has indeed changed, especially since 1945, with significant emigration to new towns after the

war and with the later immigration of sizeable ethnic minorities. This latter factor can have been of no importance in 1955, however, and of scarcely greater significance in 1965.

Immigration from the West Indies to Britain, for example, remained at a very low rate until 1955 (Peach 1968; Rose 1969), and the best estimate of the total West Indian population of the country at the beginning of that year is no more than 30 000 (Banton 1955). The large majority of these were adult males without their families. Although West Indians settled predominantly in London and other large cities, even by 1971 they formed less than 2.5 per cent of the population of London, with a maximum concentration of no more than 5 to 6 per cent in any one borough (Lee 1977). Let us take this higher figure, and suppose that Burt's 1965 sample contained 5 per cent of children of West Indian origin. If we assume that they obtained IQ scores 10 points below the white mean (Mackintosh and Mascie-Taylor 1986), this would lower the mean IQ of his sample by 0.5 IQ points. Even the most implausible and extreme assumptions about the proportion of recent immigrants in Burt's 1965 sample, and their average IQ scores, could not subtract more than one IQ point from his sample's mean score in 1965—and would still have had no effect on his 1955 scores.

There is, in fact, no reason to suppose that London schoolchildren lagged behind those in the rest of the country. In 1969, the average IQ of 11-year-olds in Inner London was slightly higher than the national average, and although the average IQ of children leaving London was marginally higher than that of those entering, this apparent net loss of people of higher IQ was substantially less than that affecting other parts of the country, such as Scotland, which has recorded significant increases in average IQ scores since 1945.* It seems extraordinarily improbable that London schoolchildren could have shown a decline in IQ at a time when children elsewhere in the country were recording significant gains. It is worth noting, indeed, that Burt himself had earlier argued that changes in IQ in London were much the same as those occurring in other large industrial towns, but that *less* favourable changes were probably occurring in rural areas of the country (Burt 1946; see below for a discussion of Burt's belief that average IQ scores were actually declining rather than increasing).

3. Burt's data might have provided a seriously inaccurate picture of changing standards of intelligence if, for whatever reason, his samples were not properly representative. As we have seen, Burt acknowledged that his figures might not be perfectly accurate, and that his samples might not all have been fully comparable with one another. Since we know virtually nothing about those samples, and nothing at all about the later

---

* The information about the IQ scores of Inner London schoolchildren in 1969 reported in this paragraph comes from an analysis of data from the National Child Development Study. I am grateful to C. G. N. Mascie-Taylor for providing me with the information.

ones, and since his method of 'median sampling' must have been capable of generating wildly inaccurate figures, it is obvious that Burt's sampling procedures might have been responsible for large discrepancies between his data and any true underlying trend. But is this sufficient to account for the actual pattern of those data? By chance, which is what bad sampling would seem to imply, Burt's data should sometimes have underestimated, but sometimes overestimated, the true level of intelligence of that particular generation of children. In other words, they would have fluctuated, rather widely, from one year to the next. But in practice, the intelligence scores in Table 5.1 show a remarkable stability, only once deviating from the 1920 mean of 100 by more than one point. And what this means is that their departure from the true underlying trend is not random, but systematic. From 1920 onward it seems almost certain that each succeeding sample increasingly *under*estimates the true test scores of a properly representative sample of London schoolchildren. How could that have happened by chance?

4. We must now consider possibilities that imply serious charges of deception or even outright fabrication. The first may sound relatively innocent, but is not. Burt's description of his testing procedures (see above) suggests that the tests used in 1955 and 1965 may have been quite different from those used in the earlier surveys. As we have seen, this would have had no consequence provided that the norms for the later tests were properly translated into those of the earlier. But perhaps they were not. Perhaps Burt relied on surveys conducted by NFER, using newly standardized tests, for 1955 and 1965, but did not translate their norms into those of his earlier revised Binet tests and standardized performance tests, described in *Mental and scholastic tests*. (Since NFER was not founded until after the Second World War, their own tests were all post-war, and they themselves did not standardize them against older norms.) As I have argued above, this would have rendered the numbers for 1955 and 1965 completely valueless, but it would serve to explain why the intelligence scores for those years continued to average about 100 (although hardly why the school attainment scores were as low as 91.4 and 93.8).

The problem with this explanation is that it would not save Burt from the charge of deliberate deception. In effect, this possibility amounts to saying that Burt relied on a procedure which he knew would produce quite meaningless data for 1955 and 1965, and that he attempted to cover this up by making the false claim that he had collected adequate new data to allow him to restandardize a 40-year-old set of intelligence tests. He explicitly states that the results from new tests 'can readily be translated into terms of the earlier norms' and that he had 'collected a good deal of fresh data' in order to restandardize his earlier tests. In fact, the norms for the 1962 edition of *Mental and scholastic tests* remained quite unchanged from those of the 1921 edition. There is no evidence that he ever collected any fresh data in 1960, in order to allow him to restandardize them.* And had he actually done so, it is inconceivable that he could have found no need to change the norms.

5. The final possibility, of course, as Hearnshaw argued, is that some of the data in Table 5.1, certainly those for 1955 and 1965, are simply fabricated. It is a rather simple matter to show that this will serve to explain many features of that table, including some that I have so far ignored.

Burt's argument, it is worth reminding ourselves, was about apparent declines in educational standards, not about intelligence at all. On the face of it, this argument would not have depended on his being able to prove that intelligence had remained stable from 1914 to 1965 (in spite of 'the vast improvement made in social conditions'). Indeed, one might suppose that if Burt were concerned to castigate progressive teaching methods, his argument would have been even more persuasive had he been able to point to a decline in educational attainments accompanied by an increase in IQ test scores. Since, as a matter of fact, there *was* a substantial increase in IQ test scores over this 50-year period, why did Burt's data not show this? The answer is because completely unequivocal evidence for such an increase was not yet available and because, along with many others of his generation, Burt had long believed that, if anything, IQ scores were *declining* from one generation to the next. The main basis for this belief was the well established observation of a negative correlation (of approximately -0.20) between IQ and family size.

In a paper entitled *Intelligence and fertility* (1946), Burt argued that this correlation, which he had himself documented in numerous surveys, meant that it must be 'the most intelligent families who contribute fewest to the next generation', and he went on to calculate that the size of the correlation implied a decline in average IQ of the population of 'about 1.5 points in a generation'—and rather more than that in rural areas (Burt 1946, pp. 17, 23). Burt was in fact being quite conservative here. On the basis of his own work, Cattell (1937) had earlier calculated a decline of about three IQ points per generation, and talked gloomily of the twilight of Western civilization, and more melodramatically of 'one of the most galloping plunges to bankruptcy that has ever occurred' (p. 43).

It is not necessary to comment on the validity of Burt's and Cattell's argument here. A negative correlation between children's IQ scores and the size of their family does not necessarily imply a decline in average IQ from one generation to the next. Burt acknowledged, as did others (e.g. Thomson 1946), that such a correlation was at best only indirect evidence: what was needed was a direct comparison of the scores obtained on the

---

* Proper restandardization of a large series of individual IQ tests (the revised Binet and the performance tests were both individual tests), for children of all ages between 3 and 16, would indeed have required something like the 'massive operation' referred to by Hearnshaw.

same tests by two representative samples of children, separated in time by ten or more years. Such data were not available in 1946, but they soon were (Cattell 1950; Scottish Council for Research in Education 1949). As we have seen, both of these studies reported small but significant *gains* in IQ scores from the 1930s to the 1940s. Burt's reaction was sceptical. In the second edition of *Intelligence and fertility* (1952), he argued that the data from the SCRE study were certainly not decisive. He noted

how slight the gain really is—less than 1 per cent on the IQ scale, a couple of the marks in the group test. (p. 43)

He talked of coaching and increased test sophistication as possible causes, and argued that if the real rate of decrease in IQ was only 1.5 points per generation,

one could hardly expect to detect it with any certainty by means of two surveys, separated by an interval of 15 years only, particularly when only a single age group was tested. (p. 44)

There were, of course, other studies appearing in Burt's lifetime, which suggested that the increases reported by SCRE and Cattell were not artefacts, lightly to be dismissed. But Burt appears to have remained unconvinced. He was not alone in this attitude. It was not until the publication of Flynn's careful and meticulous analyses of the American standardization data for the Wechsler and Stanford–Binet tests that psychometricians began to face up to the apparently inescapable conclusion that IQ test scores had increased at a remarkable rate for 50 years or more. Flynn's and Lynn's later analyses have convinced us all of the generality and reality of these increases, to the point where we may now find the earlier scepticism hard to credit. But such hindsight is misleading. As late as 1970, Burt privately dismissed, as something that could not possibly be correct, a study from New Zealand (Elley 1969) which reported a gain of nearly eight IQ points between 1936 and 1968 on the Otis test (R. Lynn, personal communication).

Burt therefore certainly refused to believe that IQ scores were generally increasing. At best, the available evidence forced him to abandon his earlier argument of a steady decline in test scores. The data shown in Table 5.1 thus accord with his own beliefs about secular trends in IQ, rather than with the facts as we now know them. But if some of these numbers were simply invented to accord with those beliefs, he was still operating under the constraints of some widely known published results. The Cattell and SCRE studies had shown a small but reliable increase in test scores from the 1930s to the 1940s. Table 5.1 also shows a small increase in IQ from 1930 to 1945—improbably small, no doubt, but not so improbable as any decrease would have been. But if Burt also wanted to establish that there had been no increase in average intelligence from 1914 to 1965, the scores from 1955 and 1965 could not be higher than 100. The solution is to make the score for 1930 particularly low.

The differences here are all very slight, and this argument will hardly persuade a sceptic. But the data on educational attainments show much larger swings, some of them extraordinarily implausible. A possible explanation is that Burt was operating under rather more stringent constraints. The Plowden Committee (1967) had established that there had been significant improvements in educational attainments since the Second World War. Burt's data also show a steady improvement from 1945 to 1965 in all educational attainment scores.* But if the 1965 scores were still to be below the 1920 baseline, the only solution was to make the 1945 scores even lower. A glance at Table 5.1 shows that indeed they are. But it seems scarcely credible that there could have been declines of 10–15 points in scores on reading accuracy, comprehension, spelling, and mechanical arithmetic between 1930 and 1945.

Conclusion

This, as Burt's defenders will be quick to note, is mere speculation, and it would be wrong to end on such a note. It is time to weigh up the various possibilities. What is the most plausible explanation of the discrepancy between Burt's published numbers and the increase in intelligence test scores which we know to have happened in Britain as elsewhere? It will be clear that I regard some combination of the final two scenarios as the most probable. An innocent explanation is hard to sustain. A ten-point gain in Britain between 1914 and 1965 is a reasonably conservative estimate. The gain *might* have been less; it is at least equally likely to have been greater. Between 1945 and 1965 alone, the increase almost certainly amounted to more than five points. There is no particular reason to suppose that Burt's tests were significantly less sensitive than others to these changes; and there is no evidence to suggest that London lagged significantly behind the rest of the country in posting these gains. It would be generous to allow these two factors to explain away more than a couple of points of the discrepancy. That leaves Burt's sampling procedure as the last (moderately) benign possibility, with the task of explaining something like an eight-point discrepancy by 1965. We can readily allow that Burt's samples may have been small and unrepresentative, and thus capable of generating random errors of this magnitude. The problem is that random error will not explain why Burt's numbers provide an ever increasing *under*estimate of the true state of affairs with each succeeding survey. This could have

---

* Joynson (1989, p. 209) argues that this implies that Burt's data were not fabricated to support his claims about declining educational standards. Had he fabricated the later data, Joynson suggests, he would have ensured that those data showed further declines. It should be obvious why he could not have done that, and why Joynson's argument is unconvincing.

happened only if Burt's samples became ever more biased towards the poor and ill-educated. And why should they have? Moreover, if they had, how could Burt have simultaneously recorded such substantial gains in educational attainments between 1945 and 1965?

At the very least then, the possibility must surely be acknowledged that Burt either relied on new tests for his post-1945 data, and that those tests had never been properly normed against his older tests, or that at least some of his numbers were simply fabricated. There are few grounds for choosing between these two scenarios—although some reason to insist that the former can hardly apply to *all* the post-1945 educational data. It hardly matters: both may be partly true, and both are equally culpable. Burt knew perfectly well (even if Miss O'Connor did not!) that newly standardized tests could provide no information about changes in intellectual or educational standards unless their norms were translated into those of earlier tests. He claimed to have undertaken the necessary translation, and to have collected a good deal of fresh data to allow him to do so. Since he did not in fact change *any* of the norms of his own tests, and since fresh data would surely have shown that he needed to, it seems highly unlikely that he ever collected any fresh data at all. Even the first of these two scenarios, therefore, has Burt making false claims in support of data which he knew would otherwise be completely worthless.

If it is impossible to decide how far Burt was relying on new but worthless data, and how far he was simply fabricating data, so some will argue, we cannot *prove* that he was doing either. That may well be true. I myself should regard some combination of these two scenarios as very much more probable than any combination of the innocent explanations I have considered. I think it would be difficult to insist that such a combination provides a *less* plausible explanation. But I readily concede that we are dealing with what is plausible or probable, not with certainty.

References

Audley, R. J. and Rawles, R. E. (1990). On a defence of Professor Sir Cyril Burt. *The Psychologist*, **3**, 306–61.

Banton, M. P. (1955). *The coloured quarter.* Jonathan Cape, London.

Burt, C. L. (1921). *Mental and scholastic tests* (1st edn). King & Son, London.

Burt, C. L. (1939). The relations of educational abilities. *British Journal of Educational Psychology*, **9**, 45–71.

Burt, C. L. (1943). Ability and income. *British Journal of Educational Psychology*, **13**, 83–98.

Burt, C. L. (1946). *Intelligence and fertility*, Eugenics Society, Occasional Papers, no. 2. Hamish Hamilton, London. (2nd edn 1952, Cassell, London.)

Burt, C. L. (1959). Class differences in general intelligence: III. *British Journal of Statistical Psychology*, **12**, 15–33.

Burt, C. L. (1969a). The mental differences between children. In *Black Paper II* (ed. C. B. Cox and A. E. Dyson), pp. 16–25. The Critical Quarterly Society, London.

Burt, C. L. (1969*b*). Intelligence and heredity: some common misconceptions. *Irish Journal of Education*, **3**, 75–94.

Cattell, R. B. (1937). *The fight for our national intelligence*. King & Son, London.

Cattell, R. B. (1950). The fate of national intelligence: test of a thirteen-year prediction. *Eugenics Review*, **42**, 136–48.

Elley, W. B. (1969). Changes in mental ability in New Zealand schoolchildren. *New Zealand Journal of Educational Studies*, **4**, 140–55.

Fletcher, R. (1991). *Science, ideology, and the media: the Cyril Burt scandal*. Transaction Publishers, New Brunswick, NJ.

Flynn, J. R. (1984). The mean IQ of Americans: massive gains 1932–78. *Psychological Bulletin*, **95**, 29–51.

Flynn, J. R. (1987). Massive IQ gains in 14 nations: what IQ tests really measure. *Psychological Bulletin*, **101**, 171–91.

Gould, S. J. (1981). *The mismeasure of man*. Norton, New York.

Hearnshaw, L. S. (1979). *Cyril Burt: psychologist*. Hodder & Stoughton, London.

Joynson, R. B. (1989). *The Burt affair*. Routledge, London.

Kamin, L. J. (1981). *Intelligence: the battle for the mind* (H. J. Eysenck versus Leon Kamin). Macmillan, London.

Lee, T. R. (1977). *Race and residence*. Clarendon Press, Oxford.

Lynn, R. and Hampson, S. (1986). The rise of national intelligence: evidence from Britain, Japan and the USA. *Personality and individual differences*, **7**, 23–32.

Lynn, R., Hampson, S. L., and Mullineux, J. C. (1987). A long-term increase in the fluid intelligence of English children. *Nature*, **328**, 797.

Lynn, R., Hampson, S., and Howden, V. (1988). The intelligence of Scottish children. *Studies in Education*, **6**, 19–25.

Mackintosh, N. J. and Mascie-Taylor, C. G. N. (1986). The IQ question. In *Personality, cognition and values* (ed. C. Bagley and G. K. Verma), pp. 77–131. Macmillan, London.

Peach, G. C. K. (1968). *West Indian migration to Britain: a social geography*. Oxford University Press, London.

Pilliner, A. G. E., Sutherland, J., and Taylor, E. G. (1960). Zero error in Moray House verbal reasoning tests. *British Journal of Educational Psychology*, **30**, 53–62.

Rawles, R. E. (1977). Correspondence. *Bulletin of the British Psychological Society*, **30**, 354.

Rose, E. J. B. (1969). *Colour and citizenship: a report on British race relations*. Oxford University Press, London.

Scottish Council for Research in Education (1949). *The trend of Scottish intelligence*. London University Press.

Sutherland, G. (1984). *Ability, merit and measurement*. Clarendon Press, Oxford.

Thomson, G. H. (1946). *The trend of national intelligence*, Eugenics Society, Occasional Papers, No. 3. Hamish Hamilton, London.

Tuddenham, R. D. (1948). Soldier intelligence in world wars I and II. *American Psychologist*, **3**, 54–6.

SIX

Burt as hero and anti-hero: a Greek tragedy

H. J. EYSENCK

Fraud or carelessness

ARISTOTLE DEFINED TRAGEDY as dramatic events which move towards a fatal or disastrous conclusion. Burt's rise and fall would certainly fit that description, and if, as Joynson (1989) and Fletcher (1991) confidently predict, a Phoenix will arise from the ashes of his fame, then his life history would fit the notion of the classical hero: Early Success through strenuous effort, Downfall through machinations of envious enemies, and final Triumph.

In this chapter I shall assume that readers are familiar with the outlines of the story, as presented by Hearnshaw (1979) and Burt's two paladins mentioned above, and will not repeat the facts and figures, true and alleged, that fill the pages of their books. The details of the charges and counter-charges are given in earlier chapters of this book. I shall instead give my personal impression of Burt, whose pupil I was through BA, and PhD apprenticeship, and in whose department I worked for a while later on. I also encountered him frequently at BPS (British Psychological Society) Council meetings, and at meetings of London University committees.

In addition I shall discuss two important matters that arise from the persecution he suffered at the hands of the media, and from the notion that if Burt cheated, then the theories of genetic determination of differences in intelligence he advocated must be wrong. And finally, I shall devote some space to a discussion of the reason why so many famous scientists, from Ptolemy to Newton, from Kepler to Mendel, and from Pasteur to Millikan, seem to have published fraudulent data (Broad and Wade 1982; Miller and Hersen 1992). It seemed possible that their historical misdemeanours might throw some light on Burt's motivation—if indeed there is anything to explain.

I will deal very summarily with what is undoubtedly the main question in most people's mind: Did Burt invent data and commit all the sins of which he stands accused? At first I was incredulous (Eysenck 1977), and voiced my incredulity; I still believe that on the strength of the material then available it would have been impossible to come to a 'guilty' verdict. Later on I became more doubtful (Eysenck 1980), until finally I reluctantly came to the conclusion that quite probably Burt had not been above-board in his dealings (Eysenck 1983). Like most people, I was very much influenced by Hearnshaw's book, but I must admit that Joynson and Fletcher have advanced many arguments that speak against Hearnshaw. Having admitted my fallibility, I myself feel that there is only one verdict possible: Not proven. This Scottish verdict means that all the evidence being circumstantial, there remains much suspicion, but in the absence of direct evidence it is impossible to come to a verdict of 'guilty'. The importance of regarding the accused as innocent until proven guilty should never be forgotten, particularly when the accused is not with us any longer, and therefore cannot defend himself.

I also think that at least some of the blame for this inability to come to any conclusion lies with the British Psychological Society. When the accusations began to appear in print, I suggested that the Society should set up a committee of independent and knowledgeable psychologists to hear evidence (most of the participants still being alive then) and come to a *reasoned* conclusion. The BPS decided to take it upon itself to clinch the issue, on the principle of 'I'll be judge, I'll be jury, said cunning old fury', and promptly condemned Burt without the precautions I had advocated. Recently they apologetically admitted their error and decided that their condemnation had been premature, but refused to do anything to try and do justice to the injured party.

The difficulty in arriving at the truth will be obvious to anyone who has read the three books in question. Fletcher and Joynson accuse many of us bystanders for not having gone through all the evidence, but few people can devote several years of their lives to such a task; we are grateful to those who have laboured on our behalf. Who can have the time or the inclination to abandon his own work and set out on such a difficult and hazardous journey? Unfortunately Hearnshaw, Joynson, and Fletcher have taken the easy way out, and have not done what I was hoping the (hypothetical) BPS committee would do, i.e. talk to all the people involved, collect information, and come to a reasonable conclusion. Two examples will show what I mean, and why I am unhappy with all three.

On p. 178 of his book, Hearnshaw refers to an 'unknown French psychologist, Jacques Lafitte', who wrote a rejoinder to a book review; Hearnshaw clearly hints that Lafitte was one of the figures dreamed up by Burt, and impossible to find, like Miss Howard. But Jacques was a fellow student of mine; he became a lecturer in Australia, wrote a book on personality (not a good one!), and died some time ago. I knew him well—indeed, I rescued his wife from suicide one day! If Hearnshaw had only bothered to ask me, I would have told him how to find the 'unknown French psychologist', who in fact was British to the core, coming from an old Huguenot family. Hearnshaw's book is full of guesses where a simple interview would have revealed the facts. Cohen (1983) gives another example of neglect. Apparently Hearnshaw did come to see him and Cohen showed him 'my high stack of Burtiana, letters, memoranda, etc., and he did not so much as suggest that he might be permitted to have a glimpse of even one of them. These unique documents clearly held no interest for him.' This is not the attitude of a conscientious historian; one does not guess where proper evidence is available. Hearnshaw never came to talk to me, and never bothered to look at my 'high stack of Burtiana'. Such failure to nail down the facts, as far as possible, makes one doubtful about the accuracy of what is revealed.

Moreover, Hearnshaw was clearly relatively innumerate, and might have benefited from taking advice from someone more knowledgeable than himself on matters of factor anlaysis, Spearman's contribution, and the question of precedence as far as the basic formula for multiple factor analysis is concerned (see Blinkhorn's chapter above). Again, I knew Spearman as well as Burt, and was familiar with the relevant literature (as well as having discussed it with Burt), but Hearnshaw never showed any inclination to discuss these matters with me. It is difficult to account for this strange aloofness.

Fletcher is certainly no better. In his book he has a chapter: 'Professors Jensen and Eysenck: questions requiring public answers'. To which the obvious reply is—why didn't he ask us? I was perfectly happy to answer any question he might like to ask, and I am sure Jensen would have done the same. As it is Fletcher makes it appear as if we were afraid to answer his queries. How, indeed, could we have given 'public answers'? What is the public in question, and who would publish the answers? The problem is that Joynson and to an even greater extent Fletcher have written their books in an *adversarial* style which is unsuited to a discovery of the truth; I suspect that they did not interview me, or Jensen, because the answers to their questions might have been inimical to their avowed points of view. The questions, in the form they are put by Fletcher, seem to accuse us of all sorts of improprieties; the answers would have taken the wind out of his sails. We once met, at the instigation of the then President of MENSA, but he never put any of these questions. 'A plague on both your houses' would seem the obvious reply to both sides. Their behaviour emphasizes the need for an *independent* committee of enquiry, willing to listen to surviving members of the scientific community who have anything relevant to say, or who have relevant written or printed material. Only in that way could obvious contradictions be resolved. Without it, and it is probably too late now that many of the people in question have died, the question of whether Burt committed fraud, or was just incredibly careless, may not be now capable of a solution. Until and unless further evidence is forthcoming, for example from a detailed examination of published papers, I do not believe that we can come to any definitive conclusion, and in the absence of such new evidence, 'not proven' remains the only justifiable verdict.

Burt the man

Even if no court of justice would find Burt guilty on the evidence submitted, all of it circumstantial, and most of it capable of different interpretations, I nevertheless believe that the character of Burt was such as to make it possible to believe that he might very well have done what he is alleged to have done. In everyday life, it is usually found that people carry on as they have done before; if we can apply a similar principle here, we may feel that perhaps facts concerning Burt's behaviour in general may be relevant to the discussion. What, after all, are character witnesses for in a court of law?

First, however, it is necessary to warn the reader that Burt's behaviour towards different people was very different. Even towards the same person he often presented a friendly face, only to try and put the dagger in from behind; Cohen (1983) testifies to Burt's agreeable side, as do Archer (1983), Banks (1983), Jensen (1983), Moore (1983), Wall (1983), and many others. Hearnshaw quotes many statements to that effect. I am happy to contribute my own testimony. In matters of academic work he was always more than ready to help, advise, and guide; a simple query would elicit typed replies sometimes amounting to several foolscap pages, with references, tables, and everything else in place. Personally, too, he was always kind, considerate, and helpful; nothing was too much for him, and he was always ready to push things along. So far, so good. But there was another side.

Hearnshaw has documented Burt's hostility towards students whom he saw as potential rivals, and I became only too aware of his obvious ambivalence towards me. He always disparaged other eminent psychologists and, as I have described elsewhere (Eysenck 1983), could not bear criticism or contradiction. He opposed my application for a chair, basing his objection on my youth and 'moral turpitude'—I had divorced my first wife and married one of my students. He apparently became so heated that the Vice-Chancellor, who presided over the Committee considering my application, stated she would be no party to a cabal! (Sir Aubrey Lewis, a member of the committee, is the source of this information.)

Burt continued his personal vendetta in many different ways. When I founded my department at the Institute of Psychiatry, he tried to reduce its range and importance by arguing, at the appropriate Board of Studies meeting, that the department should be purely for the study of abnormal psychology, that my title should be restricted in the same way, and that no work should be done there on other areas of psychology. This seemed to me both futile and silly, because there is no way in which you can prevent a professor from working on whatever research problem he wants to, and in any case it is clearly impossible to do work on abnormal subjects without having normal control groups. Fortunately, Sir Aubrey Lewis, who was the Professor of Psychiatry and the person who really created the Institute of Psychiatry in which my department was located, and who was also a member of the Board, resented the implied restrictions and argued forcibly against them. Burt, as usual, could not tolerate being worsted in argument, and the issue came up regularly at meetings of the Board of Studies, until finally everybody got tired of it. At the beginning most of the members, fearing Burt's wrath, had agreed with him, but finally the vote was cast in our favour, and the department became a department of general psychology.

The point about these examples of hostility is that they were quite uncalled for. Burt had no conceivable business to argue about the name of my department, and various members on the Board told Sir Aubrey and me that they found his behaviour odd. He went so far as to instruct his assistant S. F. Philpott, as secretary who was taking notes for the Board, to change the correct account (that the Board had rejected his suggestion) into a recommendation to consider the matter further! Several members of the Board told Lewis and me that this was typical of his insistence on always being right—he simply could not bear to be on the losing side.

Another example arose from his conduct as editor of the *British Journal of Statistical Psychology*, which he had founded. As Hearnshaw has well described, Burt wrote much of the journal himself, often under assumed names; his style is so easy to recognize that there can be no doubt of this. In itself, of course, this is not a very serious misdemeanour; most people would rather read a paper by Burt than almost any other psychologist or psychometrician; his combination of originality, rigour, and historical insight was unique, and I for one still treasure his contributions in that journal—under whatever name!

When the British Psychological Society finally decided to remove him from the editorship, he went out of his way to ensure that I would not be his successor—quite unnecessarily, as I would never have accepted the post under any circumstances! I never saw myself as a statistician, but rather as an experimentalist and theoretician who made use of statistical methods as best he could, not an innovator nor even an expert on statistical matters. I certainly did not consider myself competent to follow in his footsteps, and in any case had far too many other things to do to take up such a very time-consuming job. Burt apparently went round to different members of the Council, who later told me about this, telling them of my incompetence, and what a terrible fate awaited his cherished journal if it should ever fall into my hands! To me personally he confided how happy he would be if I were to become the editor.

Such things of course are not unusual in academic (or other!) circles. But Burt often went well beyond what is reasonable, even by the most lenient standards. Consider the following case.

One of my earliest books, *The scientific study of personality*, was being reviewed in Burt's journal (*British Journal of Statistical Psychology*, 1953, **5**, 208–12) by a well-known psychometrician, who signed himself by his initials. The review was a complete hatchet job—possibly the worst any book of mine has ever received. I am not very sensitive to criticism—either the critic is right, and this helps me to improve my theories and experiments, or else he is wrong, and can be disregarded. But this review was so biased and professionally damaging that I wrote a reply (*British Journal of Statistical Psychology*, 1953, **6**, 44–6). When I received the proof, I found that Burt had completely rewritten the reply, changing certain points and adding others I had not thought of! This seemed very odd to me, and I insisted on his printing what I had written.

I suspected that Burt had actually written the review (his style is unmistakeable), and anyone interested in seeing how his personality is mirrored in his writing should read his review and my answer; the constant deviation from accuracy and the desire to score points by misrepresentation are obvious. Actually the mysterious 'W.L.G.' who had signed the original review returned to the fray and replied to my rejoinder—using eight pages to do so. This length, plus the numerous footnotes confirmed my belief that Burt had written it all; it bore the typical stamp of his foolscap notes he sent so freely to students and others. But of course I could not be sure, so I decided to forget all about it.

Some ten years later I was standing in a queue, waiting for my tea at a BPS meeting, when someone tapped me on the shoulder. It turned out to be W.L.G., the author of the review. He told me that he had meant to talk to me many times, because the review he had written was in fact quite complimentary. Burt had complete rewritten it, introducing all sorts of gratuitous criticisms, objections, and disparagements of his own! We both agreed to find this quite unacceptable, although it also occurred to me to wonder why this man had not protested earlier—surely he must have seen the proofs? Here we have an indisputable example of very underhand dealings on the part of Burt, going well beyond the admissible.

A similar occurrence somewhat earlier left me rather bemused. I was still an undergraduate when Burt approached me with a suggestion. Thurstone had just published his original monograph on the application of his new method of multiple factor analysis to the intercorrelations between 56 intelligence tests, using a centroid method of analysis and orthogonal rotation of factors. He concluded that Spearman was wrong—there was no general factor! Burt had been asked to review this monograph for the *British Journal* *of Educational Psychology*; what he was suggesting to me was that we would both review the monograph, with him writing the text and me analysing the gigantic table of intercorrelations by means of his method of group factor analysis.* It was a gigantic undertaking this analysis of a table of over 3000 correlations, using an ancient hand-crank adding machine, but I learned a lot from it and the study is still referred to (e.g. Carroll 1993) more than 50 years later. It was of course a great honour for an undergraduate student to collaborate with his professor in this fashion, and I was happy to do so. Burt showed me his text, and told me he had sent it off with my table to the editor. When I saw it in print, I was rather surprised to see that it had appeared under *my name only*, with no mention of Burt's authorship! In addition he had rewritten the text, making it more favourable to himself and his views. These are just a few examples of how Burt made free with other people's names and contributions.

Could I have done anything? Of course not. Professor Valentine, the editor of the *British Journal of Educational Psychology*, was Burt's best friend, and how can an undergraduate argue against his professor? And would not most people believe that Burt had done me a favour by making it appear that the writing had all been mine? That I did not want largesse of that sort was immaterial. I only mention the case because it warned me of some failure of the moral sense in Burt, a failure that might not be apparent to others.

Another example of Burt's behaviour is rather more sinister. A few years after obtaining my PhD, I was working in Burt's department, in charge of some PhD students and his spokesman when statisticians and psychometricians came to the department; he tended to delegate such tasks as talking to them to his inferiors. I showed him some papers I had written on my early experimental work on humour, partly as a member of his department, partly because I genuinely valued his comments.

I was rather surprised when his only comment was: 'Eysenck, you publish too much. We don't like that in this country. There is a student of mine, Miss Schonfeld, who has been working in this field for her Ph.D., and if you publish ahead of her this will do her a lot of harm.' I could not see why it should do her any harm; presumably she was doing something quite different, and I determined to publish my articles anyway. I then got on the Underground to go home, when I saw a very attractive girl standing by the door. I chatted her up, and discovered to my amusement that this was the young lady in question. (She later got married to a Dr. Petrie, and as Senath Petrie became well known for her original work on the personality of augmenters and reducers. She was also to become one of my first PhD students.) The probability of this happening must be many times lower than that of Burt's different assessments of the intra-class correla-

---

* This may be considered approximately analogous to a Schmid–Leiman orthogonalization of eight of Thurstone's primary factors if they had been rotated to an oblique structure.

tions of his MZ twins agreeing to the third decimal, but it happened just as described.

Of course I asked her about her studies on humour, and was surprised when she replied that she was not studying humour at all! She had once toyed with the idea, and collected some cartoons, but had done nothing with them and given up the idea. How had Burt made a completed PhD out of this? I did not think about it much, being used to Burt's vagaries by then. I was brought up sharp, however, a few years later when I found a footnote in an article by Burt, on aesthetics in the journal *Character and Personality*, **7**, 1939, 291, in which he referred to Miss Schonfeld as being one among a number of his students who was doing some research on the variables he was discussing. She was not, and in fact did her PhD later on with me, on quite a different topic. In the same article he referred to experiments on smells carried out at Liverpool, the results of which (p. 293) 'have been recently confirmed by more systematic factor-analyses carried out by Mr. Roberts, Miss Davies and Mr. Eysenck.' Mr. Eysenck actually never carried out any such experiments! This kind of inaccuracy, inventing work done by people who actually did nothing of the kind, must make one cautious about his attribution of findings to other students, and constitutes a (mild) kind of fraudulence, admittedly not on a very important issue, but not something done in the best families!

I am not suggesting that it *proves* him guilty on the important point of his IQ work; it is merely an illustration of the reasons why on personal grounds I now feel it less improbable that he might have invented those data, at least in part. It is just one of many facts that show Burt less open and above-board than one likes one's scientists to be.

The media take a stand

One of the problems of dealing fairly with Burt is that the media have appropriated him as *the* exponent of views regarded by many journalists and editors as *politically incorrect*. The whole treatment of matters relating to the IQ by the media is scandalous, as Snyderman and Rothman (1988) have shown. Most of the media take it for granted that heredity has nothing to do with intelligence, that IQ tests measure nothing but ability to do IQ tests, that psychologists disagree completely about the nature and definition of intelligence, that most IQ testers regard intelligence as a 'thing' in the head, and that such a 'thing' simply does not exist. I have often argued with people holding such views (sometimes with academic qualifications like Kamin (1974), but often outside psychology), and can bear witness to their frequent ignorance of the scientific evidence, and the very theories and concepts involved.

The media regard experts like Jensen, Bouchard, Plomin, or even myself as 'mavericks', scientific outsiders who must be racists or fascists to go counter to orthodox opinion. It is noteworthy that scientific books of the

truly orthodox kind are never reviewed in the scientific or medical literature, including popular science publications like *New Scientist*, by experts
working in the field, but nearly always by outsiders with anti-psychological
attitudes and no professional knowledge of the field. Snyderman and
Rothman (1988) in their book present the views of over 600 experts on a
variety of issues in IQ testing; I found it interesting that on every relevant
issue there was agreement between the majority view and what I had said
in my textbook *The structure and measurement of intelligence* (Eysenck 1979).
Similarly, there is no disagreement between Snyderman and Rothman's
experts and the best of recent textbooks (Brody 1992). I think there can be
little doubt that all of us 'mavericks' are in fact exponents of orthodoxy,
while it is the darlings of the militant left, like Kamin (1974), who are the
mavericks.

A good example of this bias in modern media is their treatment of studies
aimed at raising intelligence. There are many such studies, claiming astonishing success—claims of 30-point improvements in IQ are not lacking
(e.g. Schmidt 1946; Shearer and Loftin 1984; Heber and Dewer 1970).
These and many similar claims for projects like 'Headstart' have been
examined by Spitz (1988) and Locurto (1991) and have been found defective. More than that, some at least were fairly clearly fraudulent. How did
the media treat the authors involved? Typically there was great publicity
and praise, often with considerable TV exposure; when the studies were
exposed, not a word appeared in the media, no retraction, nothing. In fact
some of these studies are still cited, as if nothing had happened!

The treatment accorded Burt was the exact opposite. Consider what the
BBC did to Burt in its programme 'The intelligence man'. The sub-title
describes the unfolding plot as *A story of scientific fraud*—no nonsense here
about the BBC's renowned impartiality! The accused is condemned before
the trial opens, and only witnesses for the prosecution are admitted. Both
Jensen and Cattell were interviewed but, as Cattell put it: 'We were dis-
tinctly misled.' (Fletcher 1991, p. 26). He and Jensen complain that they
were told the production would be a documentary, taking a non-judgmental
stance regarding Burt's ideas and contributions; the final product was a
hatchet job of the worst kind. I certainly failed to recognize the portrait; it
concentrated on all his bad points (even when these were purely imagin-
ary!) and left out the much larger number of good points. It assumed his
guilt, instead of fairly presenting the issues, the evidence, the difficulty of
arriving at any conclusions. As Jensen put it (Fletcher 1991, p. 26): 'This
kind of thing is typical of the popular media. I have long since learned that
the popular media (in all its forms) regard the uninspired fictions of its own
creation as more interesting than the actual truth.' Disregarding the doubtful syntax, I have to agree with the sentiment.

Fletcher (1991) gives a lengthy account of the programme, the pervasive
disharmony of its conception and presentation, the very partial atmosphere
of the whole thing, and the refusal to allow the opposite side a word in

rebuttal. Burt had his faults, but they were surely at least matched by his virtues. He may have invented some of his data, but he also made many important theoretical contributions of vital interest which should never be forgotten. His personality had faults, but which of us is without sin? Millions will have seen this deceitful and malicious programme, and formed an opinion about Burt (and about psychology!) which is a mere caricature, bearing little similarity to the man—or the profession!

And the same has to be said about the treatment of the man and the cause by the newspapers; like coyotes to the carcass they streamed in never-ending profusion, ignorant of the facts, sanctimonious to the end, *assuming guilt where evidence was doubtful or lacking*. This of course was not unexpected; the whole case was only interesting on the assumption of Burt's guilt, and accordingly this was taken for granted. The American press took up the witch-hunt, and predictably performed no better than the English press (Snyderman and Rothman 1988).

Careful perusal of what was said may give us a clue as to the reasons for the vituperative way in which the media handled the whole matter. Burt, the scientist, was of no interest to them; it was Burt the exponent of the hereditary position that interested them. Here, so it appeared, was a heaven-sent opportunity to proclaim that the position universally adopted by experts in behaviour genetics, namely that genetic factors played an important and indeed predominant part in the genesis of individual differences in intelligence, was wrong, based on the fraudulent results of one misguided extremist. That is the recurring theme of the comments made by journalists, editorial writers, commentators, and other media 'personalities', ignorant of the basic facts but very definite in knowing what they would like the outcome to be. Any stigma is good enough to strike an opposing dogma with! Snyderman and Rothman (1988) give some examples of this tendency.

The facts, of course, are quite different. In 1941, i.e. before Burt published anything on the genetic contribution to intelligence, Woodworth (1941) published his review of the evidence to date, arriving at the conclusion that the heritability of the IQ was 70 per cent, with between-family (shared) variance contributing 20 per cent, and within-family variance 10 per cent. Re-analysing the much greater amount of information then available, D. Fulker (a professional geneticist) and I came to the same conclusion (Eysenck 1979). In this analysis we omitted any data contributed by Burt, and we used the greatly improved methods of analysis then available. Obviously, the facts were quite clear; many well-executed and well-analysed genetic studies, using many different types of algorithms and methodologies, arrived at the same conclusion without making any use of Burt's contribution, and had indeed already arrived at that conclusion *before* Burt ever produced his (doubtful) results. In consequence, it is impossible to adduce his alleged fraudulence to impugn the importance of genetic factors; indeed, the improvements in methodology and analytical rigour since my earlier review (Eysenck 1979) have increased the available evidence tremendously (Brody 1992; Bouchard 1993). Burt's published figures agree very well with those published by others, both before and after he made his contribution; indeed it was because his results were so well in agreement with existing knowledge that few doubted his figures. Professor Mackintosh discusses this whole question in his final chapter.

In spite of this overwhelming agreement, the media have paid no attention; they still argue as if those of us who give expression to orthodox opinion were determined to disregard agreed facts and accept fraudulent results. When one of my friends recently suggested to a BBC producer that the inheritance of intelligence might be an interesting topic, the producer declined, saying that surely all that nonsense had been disproved! So much for the knowledge of our media experts. I have learned, in the course of bitter experience, that whenever the media discuss anything of which I have expert knowledge, what they say is more likely to be the opposite of what is true, than even resemble vaguely what has been found. Many other social scientists have had the same experience (Haslam and Bryman 1994).

Burt's importance has been grotesquely exaggerated in other ways. It is said that his writings strongly influenced American psychologists, and their attitudes to selection and educational testing. This is quite untrue: American textbooks of psychology hardly ever mentioned Burt's contribution to the genetics of IQ until they could refer to it as a fraud. Selection had its heyday *before* Burt came into the field, and has been declining ever since. American educationalists hardly ever referred to Burt. Even in this country it is quite inaccurate to picture him as the originator of the tripartite system of education, and the 11-plus as, say, Gould (1981) does. The system itself was pioneered in Germany under Bismarck, and proved eminently successful as shown by the fact that it is still in use, and that German children are achieving much higher standards than British children; in fact, only the Japanese do better, and there are reasons for doubting the application of their highly competitive methods in Europe. Burt was not a member of the Committee that introduced the 11-plus, and as a witness for the committee he argued against the 11-year dividing line, and was overruled. All this talk about his 'influence' is merely intended to paint him more deeply as the devil who frustrated egalitarian angels until his long-deserved downfall!

Does it matter? Recently I had to interview teachers, heads, and members of teachers' training colleges in connection with a long BBC programme on education. None had any knowledge of the nature or purpose of IQ testing, and there was no teaching of the principles underlying it in the Training Colleges. All that teachers, heads, and college lecturers knew was obtained from the media, and consistently wrong. The popular and completely unscientific onslaught on Burt by the media had been only too successful; by undermining the *man*, they had succeeded in also undermining the *truth* about IQ, selection, genetics, and the importance of all of these for a proper educational system. That, more than Burt's fate, is the tragedy of his case, and in so far as he has any responsibility, this is what I would find hardest to forgive. But this does not excuse his persecutors in twisting the facts, misrepresenting the position, and being economical with the truth. Two wrongs do not make a right!

Why do scientists cheat?

It is popularly supposed that, in Emerson's words, 'If a man write a better book, preach a better sermon, or make a better mouse-trap than his neighbour, tho' he build his house in the woods, the world will make a beaten path to his door.' In actual fact, of course, reviewers will slay the man who writes a better book, preachers will get their trades union to bar him who preaches better sermons, and lawyers will see to it that the man who makes a better mouse-trap is prevented from challenging the established mouse-trap makers. The same is true in science. If your wondrous discoveries offend religion, you are lucky to escape with your life (Galileo), or your reputation (Darwin). If your discoveries offend politicians, you are not likely to survive (see Medvedev 1969, 1971, concerning the fate of geneticists in Lysenko's anti-science, Stalin-supported crusade). If you offend your 'peers' in science, you may not lose your life, but you will be faced with intolerant opposition, open hostility, and a determined effort to thwart all your efforts.

Take Ignaz Philipp Semmelweis as an example. An Hungarian physician, he worked in a Viennese obstetric clinic (Slaughter 1950). His main concern was with puerperal infection, the scourge of maternity hospitals throughout Europe, which existed largely for women who gave birth in hospital because of poverty, illegitimacy, or obstetrical complication. Mortality rates for those women varied widely from time to time, and from place to place, but they reached values of 25 per cent and at times even higher. Women dreaded to be confined in hospitals. The cause of puerperal fever was unknown, and irrelevant factors such as overcrowding, poor ventilation, the onset of lactation, or simple 'miasmas' were blamed.

Semmelweis soon noted that of the two divisions of the hospital, over twice as many mothers died of childbed fever in the First Division as in the Second Division, although admission was random. The difference was that students were taught in the First Division, midwives in the Second Division. Semmelweis argued that perhaps the students carried something lethal from one patient to another, communicated when the patients were examined during labour. The students often came from an examination of cadavers, and did so without washing their hands; midwives of course did not carry out such an examination. Semmelweis was encouraged in his belief by the death of his friend, Jakob Kolletschka, a pathologist, from a wound infection. Semmelweis concluded that students who came directly from the dissecting room to the maternity ward carried infection from mothers who had died to healthy mothers they examined during labour. He promptly instituted a regime of hand-washing, using soap and water, or later chlorinated lime, prior to any examination of a woman in labour. Semmelweis later described the effect of this measure:

In 1846, when the chlorine washings were not in use, there died 459 puerperae out of 4010 in the First Division, or 11.4%. In the Second Division during 1846, out of 3754 there died 105 or 2.7%. In 1847, when about the middle of May the chlorine-washings were introduced, there had died in the First Division 176 out of 3490 puerperae, or 5.0%. In the Second Division, 32 died out of 3306 or 0.9%. In 1848, when the chlorine washings were used assiduously throughout the year, 45 puerperae died out of 3556, or 1.27%. In the Second Division during this year, 43 died out of the 3219 delivered, or 1.33%.

I have assumed that the cadaverick material adhering to the examining hand of the accoucheur is the cause of the greater mortality in the First Obstetrical Clinic; I have eliminated this factor by the introduction of the chlorine-washings. The result was, that the mortality in the First Clinic was confined within the limits of that in the Second, as the above cited figures show. The conclusion, therefore, that the cadaverick particles adhering to the hand had in reality caused the preponderant mortality in the First Clinic, was also a correct one.

Since the chlorine washings were brought into use with such striking results, there was not the slightest change made in the conditions in the First Clinic, to which could be ascribed a share in the diminution of the mortality.

An almost ten-fold reduction in mortality might have been expected to provoke praise, interest, and imitation. Nothing of the kind. The students did not like the inconvenience of constant hand-washing, and Professor Klein, his boss, driven by jealousy, ignorance, and vanity, put all sorts of obstacles in Semmelweis's way, underhandedly prevented his promotion, and finally drove him from Vienna.

This history is far closer to actual life than Emerson's imagination. History teaches us that discoveries and inventions that go against the Zeitgeist, that oppose current scientific axioms, and that have religious or social implications unwelcome to the general public will not be accepted, and may lead to all kinds of persecution. We may admit that scientists may be harrassed, maltreated, and oppressed by religious and secular leaders and bodies, but we may demur when it is suggested that other scientists may take part in such activities. Barber (1961) has documented the fact that there is a strong resistance *by scientists* to scientific discovery. Oppenheimer's (1955) book on *The Open Mind* postulated that the possession of such an open mind was almost the prerogative, and certainly the sign, of a good scientist; alas, the facts do not agree. Polanyi (1958) has emphasized the importance of the *personality* of the scientist, and no one familiar with the history of science can doubt that individual scientists, however eminent, can be as emotional, quirky, self-centred, excitable, temperamental, ardent, enthusiastic, fervent, impassioned, jealous, and hostile to competition as anyone else.

And when faced with irrational resistance, hostile incomprehension, and malicious obstruction, the creative scientist may resort to fraud—or to slight corrections of his data, to use a somewhat less brutal phrase. The books by Broad and Wade (1982) and Miller and Hersen (1992) deal mainly with frauds committed by non-entities of no importance who tried to improve their academic status and earning capacity; it is not such I am talking about. It is people of the highest stature—Newton, Kepler, Ptolemy, Mendel, and Pasteur that I am talking about; scientists whose reputation for scientific achievement could not be higher.

Let us consider the case of Newton (Westfall 1973), where the evidence seems to allow no doubt. Newton was involved in a mortal battle with continental physicists, opposing his quantitative, mathematical conception to their mechanistic views. He followed such men as Kepler and Galileo in this, but was much more concerned to establish a paradigm of 'Philosophiae Naturalis Principia Mathematica'; 'universal precision' had replaced the world of more or less, as Alexandre Koyre put it. Clearly, in such an endeavour the successful demonstration of precision was vital, and this caused Newton to fudge his data. One example must suffice, and I have chosen that of the velocity of sound. Newton had carried out the first successful analysis of what we now call simple harmonic motion, and undertook to extend that type of analysis to include the propagation of sound, thus inaugurating a new brand of theoretical physics. His demonstration rested on his understanding of the dynamics of the pendulum, which he extended to an analysis of waves on the surface of water (Westfall 1973). He demonstrated that the velocity of pulses (compressive waves propagated through air) varies as the square root of the elastic force divided by the density of water to that of air, and arrived at a preliminary velocity of sound of 979 feet per second. To this figure he made two corrections. The calculations had assumed a medium of point-like particles, but of course the real particles are of finite dimensions in comparison with the spaces between them, a property which he called the 'crassitude' of the particles, and this he used as a factor in the corrections. Another correction had to be made for vapour, which does not vibrate with the air and thus causes an increase in the velocity proportional to the square root of the amount of air that the vapour displaces. What with one thing and other, Newton's calculation finally arrived at a figure agreeing precisely with the measured velocity of sound. Unfortunately, as Westfall (1973) demonstrates, 'any number of things were wrong with the demonstration' (p. 753).

Newton assumed a precise value for the velocity of sound which he took from the average of a large number of measurements varying over a wide range. His assumption that air contains vapour in the ratio of ten parts to one, and that vapour does not participate in the sound vibration were completely arbitrary, without any empirical support. 'And his use of the "crassitude" of the air particles to raise the calculated velocity by more than 10% was nothing short of deliberate fraud' (p. 753). Newton's adjustment assumed that particles of water were completely solid; yet he believed that they contained a bare suggestion of solid matter in a vast preponderance of void! His calculation of the velocity of sound simply disregarded his quite fundamental theory of matter in order to adjust the calculated speed of sound upwards by 109 feet per second!

One can sympathise with Newton's predicament. Having nailed his colours to the mast of precision, he was left in his calculations with an uncovered discrepancy of 21 per cent, giving priceless ammunition to the hostile continental physicist jeering at him! 'The very flagrance of his adjustment in this case becomes evidence for the compulsion behind the pretence of precision in the other case' (p. 754), i.e. the acceleration of gravity, and the precession of the equinoxes, where he similarly fudged the data.

Kepler, too, presents a clear case of fraud (Donahue 1988). His book, *New astronomy*, published in 1609, used the results of triangulation in discovering the shape of the orbit of Mars, but only as a guide to his theorizing (Wilson 1968). Wilson argues that Kepler could not possibly have used triangulation to determine the orbit, because the procedure of triangulation was too imprecise. Kepler's presentation of his data is muddled and contradictory. The promised triangulation is reported in Chapter 51, but instead of proceeding directly to the comparison of the circular and eliptical theories, Kepler returns to the same question in Chapter 53, in a curious hybrid procedure difficult to follow. Donahue has this to say about Chapter 53:

A closer look at Chapter 53, far from answering this question, only serves to increase the perplexity. After telling us what he is going to do, Kepler proceeds to give two purported examples, which actually show a quite different procedure. Then, excusing himself from further presentation of computational details on the grounds that 'it would be tedious', he sets out the table mentioned above. But, we are startled to note, the numbers in the table are dramatically different from the numbers in the computations from which they were supposedly derived. And that is not all: at the beginning of the next chapter, Kepler refers to a fifteen-minute correction in the mean anomalies, an astonishingly large adjustment that he claimed was introduced in Chapter 53. Nothing of the kind is to be found in that chapter, however. Clearly there is more here than meets the eye. (Donahue 1988, p. 217)

It is certainly startling to find an absence of essential computational details because 'taediesum esset' to give them. But worse is to follow.

Donahue makes it clear that Kepler presented *theoretical deduction* as *computations based upon observation* (shades of Dorfman's (1978) charge against Burt). He appears to have argued that induction does not suffice to generate true theories, and to have substituted for actual observations figures deduced from the theory. This is historically interesting in throwing much light on the origins of scientific theories, but is certainly not a procedure recommended to experimental psychologists by their teachers.

A good modern example of fraud is the case of Freud (Eysenck 1985). Freud consistently told lies about his most famous cases, like the Wolf Man (Obholzer 1982), the Rat Man (Mahony 1986), and Anne O. (Thornton 1983). Scharnberg (1993) has documented many cases of 'The non-authentic nature of Freud's observations', and his book should be consulted by anyone still doubtful about the fraudulent nature of many of Freud's observations. What is particularly interesting is not Freud's misrepresentations of the facts, but the complete failure of the majority of psychologists and psychiatrists to take note, and dismiss his claims outright for lack of sound evidence.

Many people have difficulties in understanding how a scientist can fraudulently 'fudge' his data in this fashion. The line of descent seems fairly clear. Scientists have extremely high motivation to succeed in discovering the truth; their finest and most original discoveries are rejected by the vulgar mediocrities filling the ranks of orthodoxy. They are convinced that they have found the right answer; Newton believed it had been vouchsaved him by God, who explicitly wanted him to preach the gospel of divine truth. The figures do not quite fit, so why not fudge them a little bit to confound the infidels and unbelievers? Usually the genius is right, of course (if he were not, we should not regard him as a genius), and we may in retrospect excuse his childish games, but clearly this cannot be regarded as a licence for non-geniuses to foist their absurd beliefs on us. Freud is a good example of someone who improved his clinical findings with little regard for facts (Eysenck 1985).

Such, then, are the conditions in which leading scientists may be led to falsify their data, or invent them. Convinced (rightly as it happens) that they have made a stupendous discovery, of the greatest scientific and social importance, they see that discovery threatened by envious or hostile people with the power to destroy them. The first findings in trying to substantiate a theoretical discovery are never clear cut; there were anomalies in Newton's laws for 300 years, all of them capable of being incorporated in his system (until the beginning of this century!). But enemies would seize upon these anomalies to destroy his theory. Obviously the way to overcome this problem is simply to make sure the data fit the theory! If this is indeed what happened, then it seems possible that the same happened in Burt's case—always assuming that he did invent some of his data.

Burt was firmly convinced of the *truth* of his theories—and indeed had every reason to be sure. But by the 1950s and 1960s, the Zeitgeist began to turn against him. The Labour government of the 1960s demolished one of the best educational systems in Europe and, in the name of 'equality', abandoned selection and opted for 'comprehensive schools', with the result (predicted by Burt and many others in a series of 'Black Papers on Education') that British education slumped to the bottom of the European league, with achievement quotients two years behind Germany, France and other European countries (Naylor 1985; Cox and Marks 1988). Even some of the authors of this policy have recognized its failure, and urged reconsideration. Burt saw his cherished dreams of educational practices being governed by scientific laws of learning and intelligence destroyed by politicians, sociologists, and educational theorists ignorant of the facts; perhaps he thought that it was his duty to combat this self-defeating rush of the lemmings, and tried to shore up his campaign with invented data? We can never be sure of what he did, and even less of why he did it, but this guess may not be entirely mistaken.

I used to think that the best argument for Burt's innocence was the failure on his part to disguise whatever frauds he might have committed. He was a genius in the fields of statistics and psychometrics, he could easily have avoided leaving any traces of wrong-doing. But the same is true of Newton and Kepler, and their misdeeds are only too obvious. We might ask why no one until recently pilloried these historical figures, or the less historical Freud, when there was really never any doubt about their malfeasances? Burt, admittedly, is a much lesser figure; is there one law for geniuses, another for the less eminent? Newton, Kepler, Mendel, Pasteur, and Ptolemy were right in the essentials of what they wrote; perhaps it is the winners who write (scientific) history? But Freud was a loser, as we now know, and yet he has never been called to account. I do not know what the solution to this question may be, nor whether the future will eventually give a more equitable verdict on Burt. 'The evil that men do lives after them, the good is oft interred with their bones.' Perhaps it is time to resurrect the good.

Conclusion

From my intimate knowledge of Sir Cyril Burt, I have suggested that in addition to many good personality traits he also had some qualities that were not attractive; in this he was like most of us. His intellectual qualities were outstanding; at his best there were few, if any, to equal him. But at bottom he preferred theory and statistical analysis to experimental rigour and hypothesis testing along deductive lines. No certain verdict is possible on the allegation that he committed fraud; in view of such evidence as there is, 'not proven' seems the most appropriate judgment. But most of his findings, whether factual or imagined, are well in line with earlier and later work; even if he is found guilty it would not make the slightest difference to the theory he put forward. We know he was right, and he certainly made notable contributions to the theory of intelligence, of behavioural genetics, and of education. His tragedy was to offend against 'political correctness', and to be chosen as the major target for egalitarians who could not tolerate the variety of talents nature bestows upon us.

References

Archer, G. (1983). Reflections on Sir Cyril Burt. *Journal of the Association of Educational Psychologists*, **6**, 53–5.

Banks, C. (1983). Professor Sir Cyril Burt: selective recollections. *Journal of the Association of Educational Psychologists*, **6**, 21–42.

Barber, B. (1961). Resistance by scientists to scientific discovery. *Science*, **134**, 596–602.

Bouchard, T. (1993). The genetic architecture of human intelligence. *Biological approaches to the study of human intelligence* (ed. P. A. Vernon), pp. 33–94. Ablex, Norwood, NJ.

Broad, W. and Wade, N. (1982). *Betrayers of the truth: fraud and deceit in science.* Oxford University Press.

Brody, N. (1992). *Intelligence.* Academic Press, New York.

Carroll, J. (1993). *Human cognitive abilities.* Cambridge University Press.

Cohen, J. (1983). Sir Cyril Burt: a brief note. *Journal of the Association of Educational Psychologists*, **6**, 64–77.

Cox, C. and Marks, J. (1988). *The insolence of office.* Claridge Press, London.

Donahue, W. H. (1988). Kepler's fabricated figures. *Journal of the History of Astronomy*, **19**, 217–237.

Dorfman, D. (1978). The Cyril Burt question: new findings. *Science*, **201**, 1117–86.

Eysenck, H. J. (1977). The case of Sir Cyril Burt. *Encounter*, **48**, 14–24.

Eysenck, H. J. (1979). *The structure and measurement of intelligence.* Springer Verlag, New York.

Eysenck, H. J. (1980). Psychology of the scientist: XLIV. Sir Cyril Burt: prominence versus personality. *Psychological Reports*, **46**, 893–4.

Eysenck, H. J. (1983). Sir Cyril Burt: polymath and psychopath. *Journal of the Association of Educational Psychologists*, **6**, 57–63.

Eysenck, H. J. (1985). *Decline and fall of the Freudian empire.* Viking, London.

Fletcher, R. (1991). *Science, ideology and the media: the Cyril Burt scandal.* Transaction Publishers, London.

Gould, S. J. (1981). *The mismeasure of man.* Norton, New York.

Haslam, C. and Bryman, A. (1994). *Social scientists meet the media.* Routledge, London.

Hearnshaw, L. (1979). *Cyril Burt: psychologist.* Hodder & Stoughton, London.

Heber, R. F. and Dewer, R. B. (1970). Research on education and habilitation of the mentally retarded. In *Social-cultural aspects of mental retardation* (ed. H. C. Haywood), pp. 194–216. Appleton-Century-Crofts, New York.

Jensen, A. R. (1983). Sir Cyril Burt: a personal recollection. *Journal of the Association of Educational Psychologists*, **6**, 13–20.

Joynson, R. B. (1989). *The Burt affair.* Routledge, London.

Kamin, L. J. (1974). *The science and politics of IQ*. Erlbaum, Potomac, MD.

Locurto, C. (1991). *Sense and nonsense about IQ*. Praeger, New York.

Mahony, P. J. (1986). *Freud and the rat man*. Yale University Press, New Haven, CT.

Medvedev, Z. A. (1969). *The Rise and Fall of T.D. Lysenkov*. Columbia University Press, New York.

Medvedev, Z. A. (1971). *The Medvedev Papers*. Macmillan, London.

Miller, J. and Hersen, M. (1992). *Research fraud in the behavioral and biomedical sciences*. Wiley, New York.

Moore, T. (1983). Thoughts on the integrity of Sir Cyril Burt. *Journal of the Association of Educational Psychologists*, **6**, 42.

Naylor, F. (1985). *Technical Schools: A tale of four countries*. Centre for Policy Studies, London.

Obholzer, K. (1982). *The wolf man: sixty years later*. Routledge & Kegan Paul, London.

Oppenheimer, R. (1955). *The Open Mind*. Simon & Schuster, New York.

Polanyi, M. (1958). *Personal Knowledge*. Routledge & Kegan Paul, London.

Scharnberg, M. (1993). *The non-authentic nature of Freud's observations*. Acta Universitatis Upsaliensis, Uppsala.

Schmidt, B. G. (1946). Changes in personal, social, and intellectual behaviour of children originally classified as feeble-minded. *Psychological Monographs*, **60**, (5), Serial No. 281.

Shearer, D. E. and Loftin, C. R. (1984). The portage Project. In *Parent training* (ed. R. F. Dangel and R. A. Polster), pp. 95–118. Guilford, New York.

Slaughter, F. G. (1950). *Immortal Magyar: Semmelweis, Conqueror of Childbed Fever*. H. Wolff, New York.

Snyderman, M. and Rothman, S. (1988). *The IQ controversy*. Transaction Books, Oxford.

Spitz, H. H. (1980). *The raising of intelligence*. Erlbaum, London.

Thornton, E. M. (1983). *Freud and cocaine: the Freudian fallacy*. Bland & Briggs, London.

Wall, W. D. (1983). Sir Cyril Burt: a personal note. *Journal of the Association of Educational Psychologists*, **6**, 43–5.

Wilkinson, M. (1977). *Lessons from Europe*. Centre for Policy Studies, London.

Wilson, C. (1968). Kepler's derivation of the elliptical path. *Isis*, **59**, 75–90.

Woodworth, R. S. (1941). *Heredity and environment*, Bulletin 47. Social Science Council, New York.

SEVEN

Does it matter? The scientific and political impact of Burt's work

N. J. MACKINTOSH

UNTIL THE SAGA of Burt's possible fraud began to unfold, the most celebrated, and least equivocal, case of British scientific fraud in the twentieth century was the Piltdown Man hoax. The fragments of skull, jawbone, and teeth discovered by Charles Dawson at Piltdown in Sussex between 1908 and 1915 appeared to provide evidence of the missing link between ape and human required by the theory of evolution. The hoax was extraordinarily successful, for it was not until 1953 that Piltdown Man was formally declared to be a fake, the jawbone that of a modern ape, the skull and teeth human (the latter filed down). But the chief measure of the forgery's success was that it led so many British (and other) anthropologists down the wrong path for some 30–40 years. If early hominids had really developed a modern size brain while still retaining many ape-like features (the reverse of the truth as we now understand it), then, for example, the australopithecine fossils discovered in South Africa by Raymond Dart and Robert Broom in the 1920s and 1930s could have had nothing to do with the story of human evolution.

Is there any parallel here with the case of Cyril Burt? Did the (possibly false) data he published affect the course of science? Even if we give Burt the benefit of every possible doubt, it is no longer possible to accept at their face value the data presented in his later papers (e.g. Burt 1961, 1966, 1969). Giving him that benefit would mean accepting that he or his assistants did actually collect some information about the IQ scores of 53 pairs of separated MZ twins, of some 1000–2000 fathers and their sons, and of successive generations of London schoolchildren between 1914 and 1965. But his accounts of these data are so woefully inadequate and riddled with error, and some of the data must be based on such grossly inadequate methods, that no reliance can be placed on the numbers he presents. They are, indeed, as Kamin said they were, not worthy of our current scientific attention. But they were, of course, widely accepted, and figured prominently in numerous books and reviews. Arthur Jensen, in words he must have lived to regret, described Burt's work as 'the most satisfactory attempt' to estimate the heritability of IQ (Jensen 1969). Hans Eysenck was no less complimentary:

The long-continued studies of Burt have been particularly valuable in throwing light on the relation between IQ and social class. I shall draw rather heavily on his work . . . because of the outstanding quality of the design and statistical treatment in his studies. (Eysenck 1973, p. 120)

Does this mean then that Burt's data, like the Piltdown hoax, perverted or misdirected the course of scientific enquiry?

Burt's critics, of course, have long argued that his so-called data not only perverted the course of science, they also had a malign influence on social and educational policy. Indeed, one of their more common, if not always explicit, claims has been that Burt invented his data precisely in order to provide support for a variety of malign policies. Thus the attack on Burt has often been an attack on his reactionary social and political views. Oliver Gillie's *Sunday Times* article (1976) began by stating that

leading scientists are convinced that Burt published false data and invented crucial facts to support his controversial theory that intelligence is largely inherited.

In November 1978, *The New York Times* characterized Burt as a 'long-time advocate of the genetic basis of racial differences in intelligence'. The BBC TV programme, 'The intelligence man', depicted Burt as a man sympathetic to eugenic ideas, contemptuous of Jews, communists, and slum children. Nor were such attacks confined to newspapers and television. Leon Kamin's attack on Burt was launched in a book whose central thesis was that

IQ tests have been fostered by men committed to a particular social view. That view includes the belief that those on the bottom are genetically inferior victims of their own immutable defects. The consequence has been that the IQ test has served as an instrument of oppression against the poor—dressed in the trappings of science, rather than politics. (Kamin 1974, pp. 1–2)

Kamin's attack on Burt's science was liberally interspersed with denunciations of his social and political prejudices, and although he did not at that point accuse him of outright fraud, he later wrote:

The clear implication—that Burt had invented the data in order to support his ideas about social and educational policy—was left for the reader to make. (Kamin 1981, p. 102)

It would be idle to pretend that psychologists' social and political views have had no impact on their attitude towards IQ tests (although Burt's own views seem to have borne rather little relation to those foisted on him by some of his attackers). One trouble with Kamin's argument, however, is that it can be directed at Kamin as readily as against Burt. If we are asked to question Burt's science because it may have been influenced by his political prejudices, why should we not question Kamin's science for exactly the same reason? Indeed, the inevitable consequence of the type of attack which Kamin initiated, and which was taken up by Gillie and others, has been a backlash of support for Burt, based on the claim that the attack was motivated by politics rather than any quest for scientific truth. The very title of Fletcher's defence of Burt, *Science, ideology and the media* (1991) accurately reflects his main line of argument: that the attack on Burt was ideologically motivated and taken up by the media in a lurid and irresponsible manner. Others have not been slow to join in. Philippe Rushton (1994), for example, has argued that it was Burt who was the victim of a hoax—the hoax of 'genetic equalitarianism', which denies the possibility of inherited differences in mental capacity. According to this defence, the best reason for believing Burt innocent of the charge of fraud is that the charge was motivated solely by political prejudice.

Kamin wrote that his book was

about the politics of intelligence testing, as well as the science of intelligence testing. To pretend that the two are separable is either naive or dissembling. (op. cit., p. 2)

At the risk of appearing both, I believe it is worth trying to return to the scientific questions. What was the scientific impact of Burt's work? Did the general acceptance of his data have a deleterious effect on the course of scientific enquiry? In so far as we can now judge it, what is the truth about the various scientific issues on which Burt pronounced? This book has not touched on all aspects of Burt's work, and it certainly lays no claim to provide a history of British psychology in the first half of the twentieth century. I shall confine my discussion to those topics and issues which have been addressed in earlier chapters.

Factor analysis

Burt's major theoretical contribution to psychology was surely his work on the development of factor analysis. No one has ever disputed that Spearman was the first to see that the positive correlation between a variety of mental tests was consistent with the possibility that performance on all such tests was dependent on a single factor of general intelligence, or that he developed techniques for analysing such intercorrelations. But as Blinkhorn, I believe, shows, it was Burt, not Spearman, who actually developed the technique of 'simple summation', the precursor of modern factor analysis, which he derived from Pearson, to apply to these problems. Unquestionably, it was Burt who was the first to see that the actual pattern of correlations observed between a large battery of intelligence tests could not be explained simply in terms of Spearman's general factor, but implied the existence of other 'group' factors. Indeed, the techniques which Spearman employed could only be applied to a battery of tests dependent on a single, general factor, and thus could not determine the nature of any group factors. What Hearnshaw saw as a rewriting of the history of factor analysis after Spearman was safely dead and buried is more charitably and reasonably understood as Burt's justifiable resentment that his earlier contribution had been wholly ignored by Thurstone. Whether or not Hearnshaw was 'pretty innumerate', as Eysenck supposes, he was insufficiently conversant with the mathematics of factor analysis to understand that what Spearman did to substantiate his two-factor theory of intelligence was quite different from the procedures developed by Burt and later by Thurstone. It is thus, in a sense, misleading to call Spearman the father of modern factor analysis in psychology. In one sense, at least, that title really does more properly belong to Burt. Moreover, Burt's insistence that factor analysis yields evidence both of a general factor and of several group factors has stood the test of time a great deal better than either Spearman's insistence on only a general factor or Thurstone's attempt to get rid of it. Of course, factor analysis has not been without its critics. Properly understood, however, it remains an important and useful technique—even if it will never reveal the nature of the psychological process or processes responsible for the appearance of the general or group factors that the technique reveals. That, being a psychological question, requires the methods of psychology, not statistical techniques, for its answer.

Social class and IQ

Hearnshaw described Burt's (1961) paper, 'Intelligence and social mobility', as 'contentious'. Of course, the whole issue of social class differences in IQ has been a contentious one, and many critics have denounced IQ tests, and those who devise them, as biased precisely because they appear to reveal differences between the average scores obtained by different social, cultural, or ethnic groups. Burt's paper was widely cited, however, not because it was the first to provide evidence on this topic, but because at first sight (and with the advantage of hindsight, we can now say, on a casual and careless reading), the data he presented seemed unusually clear and elegant, and the argument he developed particularly persuasive. On the central issues, however, his data, whether fabricated or not, are broadly consistent with those of others. That there are substantial differences in average IQ between children from different social classes was amply documented in Terman's original standardization of the Stanford–Binet test (Terman 1916). Since then, social class differences in average IQ have been one of the best documented findings in the IQ literature (see, for example, Bouchard and Segal 1985; Jencks 1972; Rutter and Madge 1976; White 1982). The differences are found in both adults and children, but are larger in the former: the correlation between adult IQ and social class is

0.50 or higher; that between children's IQ and the social class of their parents is only about 0.30. Unless average social class differences in IQ are disappearing (and there is no evidence that they are) it follows, as Burt argued, that there must be some tendency for the difference to be recreated in each generation. Studies in both Britain and the US confirmed that social mobility may be correlated with IQ: sons with higher IQ than their fathers were more likely to end up in higher-status jobs; those with lower IQ were more likely to move down in social class (Mascie-Taylor and Gibson 1978; Waller 1971).

Of course, since Burt's (1961) paper was published, there has been a considerable body of new research on social class and social mobility, not only in Britain and the USA but also elsewhere. For example, Goldthorpe devised a seven-point class scheme, based on the notion of 'the market and work situations', which differs from the standard Registrar-General's classification in terms of both its theoretical rationale and arrangement of classes (Goldthorpe and Hope 1974; Goldthorpe *et al.* 1980). The Oxford Mobility Study (Halsey *et al.* 1980) examined the social origins and educational destinations of a large representative sample of adult males in England and Wales, and related mobility to differences in family background and education, as well as IQ. They used a path analysis to assess the relative contribution of various factors, concluding that IQ was less important than others. Whether this is the last word on the matter need not concern us here. We can readily allow that many social scientists today would disagree with some of Burt's arguments and conclusions, some of which have already been overturned by more sophisticated analyses. But whether or not his data were fabricated, they were not seriously at variance with what other investigators were saying at the time. Waller (1971), for example, acknowledged the extent to which his work simply confirmed what Burt had said. In that sense, Burt's work did not pervert or misdirect the course of scientific enquiry.

Declining educational standards

Burt (1969) argued that improvements in education over the 50 years since the First World War had been more apparent than real, and that important, basic educational standards had declined over this period. I do not think it worth spending much time discussing this claim. The data Burt presented to support it agreed with other studies in showing that there had been significant improvements between 1945 and 1965, but in order to reconcile this with an overall decline from 1914 to 1965, his data also appeared to show a quite remarkable decline in standards between 1930 and 1945. I know of no other data consistent with this. But the real problem here is to see how such changes could ever be sensibly measured. What children were taught in school, and the way they were taught it, surely changed between 1914 and 1965. Tests geared to 1914-style teaching might reveal an apparent decline in performance by 1965. Tests geared to 1965-style teaching might have shown an improvement. Which is the more valid conclusion?

We are on more secure ground when we turn to Burt's arguments and data on intelligence in his 1969 paper. Along with others of his generation, Burt believed that the negative correlation between children's IQ scores and the number of other children in their family implied that less intelligent parents were having more children than those who were more intelligent. From this he concluded that the average intelligence of the population as a whole must be declining. He continued to maintain this conclusion as late as his paper on social mobility, where he stated that there had been a decline of one or two points per generation (Burt 1961), but he did begin to backtrack in the face of a steady trickle of evidence to the contrary. By 1969, his data now showed no more than a one-point decline between 1914 and 1969. But this was sufficient to imply that intelligence was largely inherited, since it had failed to increase in spite of substantial improvements in social and, supposedly, educational circumstances.

Although at one time it appeared otherwise, there is now good evidence that, in the US at least, there has been a small negative relationship between parental IQ and fertility for most of the twentieth century (Van Court and Bean 1985). Thus Burt may well have been right on this score. But he was quite wrong in the conclusion he drew. We now know that performance on IQ tests has been improving almost from the day the tests were first introduced. Indeed, it is the conflict between Burt's data and what we now know to be the truth that provides the strongest grounds for believing that Burt's data are fraudulent. He must either simply have fabricated the data, or have relied on different tests at different times, which, in spite of his explicit claims to the contrary, he had never standardized against each other.

The consequences of Burt's fraud here, however, were wholly insignificant. The evidence that IQ test performance had increased substantially since 1914 was already beginning to come in, and it was not long before this conclusion was securely established (Flynn 1984, 1987) and generally accepted. No one paid any attention to Burt's apparently contradictory evidence, which was published in a relatively obscure journal. Indeed, as far as I know, no one until now had noticed the contradiction, or drawn the conclusion that Burt's data are therefore suspect.

The heritability of IQ

The main issue taken up by Burt's critics, of course, was his 'controversial theory that intelligence is largely inherited'. Burt argued that his kinship data, in particular his data on separated identical (MZa) twins, established that individual differences in 'intelligence', particularly when the assessments have been carefully checked, are influenced far more by genetic constitution, or what is popularly termed 'heredity', than by post-natal or environmental conditions. (Burt 1966, pp. 151–2)

If those data were fraudulent, or even if just unreliable, some critics argued, the theory too could be dismissed. This was the picture typically painted in the media. Burt's controversial theory was just that—*his* theory, shared only by a handful of fellow reactionaries; it was obvious, therefore, that it must rest on his data, and equally obvious that it must fall with them. As will be clear from their chapters above, both Jensen and Eysenck categorically reject this argument, insisting first that Burt's own views were, and still are, shared by all experts in the field, and secondly that the loss of his data makes no difference to the precise estimate of the heritability of IQ, or to the confidence we can place in such an estimate.

Their first point is certainly nearer the truth than the critics' claim that a belief in the heritability of IQ is confined to a small number of ignorant and reactionary zealots. Standard textbooks of psychology have taken it as an established fact for the past 50 years or more. As Eysenck notes above, Woodworth concluded in 1941 that the heritability of IQ was probably approximately 0.70. In a more recent survey of some 650 social scientists, with credentials in developmental and educational psychology or psychological testing, Snyderman and Rothman (1988) reported that of the 75 per cent who felt qualified to answer the question, the overwhelming majority thought that IQ had significant heritability.

There can be no doubt that this is the consensus view among those familiar with the evidence. But that does not *necessarily* mean that it is correct. And it certainly does not mean, as Jensen and Eysenck have implied, that the loss of Burt's data never had any consequence for the hereditarian case. Indeed it is hard to see how that could have been true. If Burt's data really had appeared to provide the best and most systematic evidence on the heritability of IQ, their loss ought to have had some impact on our estimates of heritability. I believe that it did.

Comparison of Burt's data with other studies of MZa twins
One of the less convincing arguments advanced by both Joynson (1989) and Fletcher (1991) to prove that Burt's data were genuine is that those data agree so closely with data reported by others. The sceptic is justified in replying that the mark of a successful scientific fraud is precisely that he manufacture data not too seriously at variance with what is generally believed. The Piltdown hoax was successful precisely because some anthropologists wanted to believe that the development of the brain led the way in human evolution. Moreover, the agreement between Burt's data and others' is not quite as close as has sometimes been suggested. The

Table 7.1 IQ correlations for MZa twins

Study	Test		
	Individual	Group	Final assessment
Newman, Freeman, and Holzinger (N=19)	0.67	0.73	—
Shields (N=38)	—	0.77	—
Juel-Nielsen (N=12)	0.68	0.77	—
Burt (N=53)	0.863	0.771	0.874

fact of the matter is that Burt's separated identical twins resembled one another rather more closely (thus implying a somewhat higher heritability for IQ) than did the twins in the other three studies reported in his lifetime. The relevant data are shown in Table 7.1. In all of these studies, the twins were given more than one IQ test—hence the two different scores that appear for Newman *et al.* (1937) and Juel-Nielsen (1965). Shields's (1962) two tests have usually been combined to yield a single score. In Burt's case, in addition to the individual and group tests, there was also his 'final assessment'—deliberately intended to make allowance for large discrepancies in environmental experience, and thus not strictly comparable with the IQ scores published in other studies. Although Burt's group test gives a correlation comparable with those reported in other studies, the results of his individual test, which gave virtually the same correlation as his final assessment, are well outside the range of the other studies.

These discrepancies should not be exaggerated. Burt's data are not *greatly* at variance with those reported in the other three studies, and Jensen (1972) was able to show that there was no difference between the four studies in the size of the differences in IQ score between one twin and another within each pair. To this extent, Jensen and Eysenck seem justified in claiming that the loss of Burt's data did not *greatly* matter. But it is not for nothing that they also argued that Burt's data were the best available. For purposes of estimating the heritability of IQ, they were incomparably better than those of the remaining three studies.

Estimating the heritability of IQ
To see why, we need to understand the logic of these MZ twin studies— indeed we need to go back far enough to understand how any kinship correlation can be used to estimate the heritability of IQ. In outline, at least, the argument is reasonably simple. If you give an IQ test to randomly chosen, unrelated pairs of people, you will find, unsurprisingly, that they do not resemble one another very closely in IQ. Indeed, with a large enough sample, you will certainly find that the correlation between one

member of each pair and the other is roughly zero. This is unsurprising, since if they have been chosen at random, the members of each pair will neither be related to one another, nor will they have shared particular environmental circumstances in common. Neither shared genes nor shared environment is working to produce any resemblance in IQ. Let us now go to the other extreme: MZ twins, brought up together, resemble one another very closely in IQ—their correlation being somewhere between 0.80 and 0.90, which is not much less than the correlation between your IQ score today and your score in a week's time: IQ tests do not give *precisely* the same score on every occasion. If MZ twins brought up together resemble one another so closely, this is presumably because they share in common most of the factors that influence IQ scores. And of course they do: they share the same set of genes—and have also experienced much the same environment. The problem, of course, is to disentangle these two sources of resemblance. MZ twins brought up apart seem to provide the answer. They too are genetically identical, but if they really have been brought up in different environments and had no contact with one another, their resemblance must be attributed *solely* to their shared genes.

But here is the catch. Social scientists, and human behaviour geneticists, must rely on naturally occurring experiments, and these are necessarily imperfect. Social scientists cannot take 50 pairs of MZ twins from their mothers at birth and assign them, at random, to a variety of different homes. They must rely on those rare cases where the parents of such twins feel unable to bring up both themselves, and choose to give up one or both for adoption. These parents' choices will, no doubt, be determined by a variety of different factors, but of one thing we may be reasonably confident: the last thing on their mind will be the design of a perfect adoption study for the sake of the behaviour geneticist. After Burt's, the largest of the three studies in Table 7.1 is that of Shields. Here, the most common pattern, true of 27 of his 38 cases, was for one twin to be brought up by the natural parents and the other by a relative (usually an aunt or grandmother). Relatively few of the twins had been separated at birth; one pair had lived together until the age of nine; others were reunited for varying lengths of time in childhood; several lived in the same village or town, and some attended the same school. How can we be sure that it is not these shared experiences and similar family environments that is largely responsible for their resemblance in IQ?

Burt implied that his own study did not suffer from these problems. Although providing no data on their degree of contact after they had been separated, he stated that all his twins had been separated before they were six months old (Burt 1966, p. 141), and that only in five pairs had both twins been brought up by relatives. But in each of these cases, one 'lived in a town and the other in the country' (op. cit., p. 143). Finally, he presented data to show that there was no correlation between the socio-economic circumstances of the two twins' homes, and that these circumstances were reasonably representative of those to be found in society as a whole (op. cit., Table 1; see above, Chapter 3, Table 3.6, for the final form in which these data appeared).

The importance of these features of Burt's data can hardly be exaggerated. If all this were true, the implication is that the twins really did experience rather different environments, and the only plausible explanation for the resemblance between their IQ scores would be their shared genes. The correlation between these IQ scores would provide us with a virtually direct and uncontaminated estimate of the heritability of IQ.

On the face of it, then, it is scarcely credible to suggest that the loss of Burt's data would have no impact on estimates of the heritability of IQ, or on the confidence which could be placed on those estimates. According to Kamin, at least, and to others who followed in his footsteps, the remaining three twin studies shown in Table 7.1 were so full of flaws, the twins so manifestly not properly separated, that no unequivocal conclusion could be drawn from them. Without Burt, so the argument ran, the evidence for the heritability of IQ looked decidedly unimpressive.

Taylor (1980) and Farber (1981) have written entire books devoted to an analysis of these three twin studies: rarely can 138 individuals have been the subject of such detailed, painstaking, and nitpicking analysis. The central point on which all these critics agreed is that the resemblance between the twins was partly a consequence of imperfect separation. When Kamin divided up Shields' sample into those brought up in related families and those brought up in unrelated families, he found that the former showed a correlation of 0.83, and the latter a correlation of only 0.51. Kamin also calculated the correlation for seven pairs

selected as the most striking examples of correlated environments in Shields' sample . . . The intelligence correlation is .99. That seems incredible. Tests are never that reliable, nor are *nonseparated* twins so highly correlated. (1974, p. 51)

This correlation contrasted with one of only 0.47 for the ten pairs who had neither attended the same school nor been brought up in related families. He concluded that there was nothing in Shields' data to contradict the assumption that, with truly uncorrelated environments, MZ twins would show no correlation for IQ.

Taylor, analysing the data from all three studies together, used four criteria of degree of shared environment: age at separation; reunion in childhood; whether the twins were brought up in related or unrelated families; the similarity of their families' social circumstances. He reported that all but the first of these factors had a profound effect on the resemblance between twins' IQ scores. When twins were reunited in childhood, lived in related families, or in similar social circumstance, their IQ scores correlated about 0.80. In the absence of these sources of similarity, the correlations dropped to 0.50 or so. When he took the five pairs of twins from Shields' study who, he argued, had clearly experienced quite different upbringings, the correlation in their IQ scores was only 0.24. Here was proof of Kamin's conjecture. The conclusion seemed inescapable: once Burt's data were removed from the equation, separated MZ twins provided virtually no evidence for the heritability of IQ.

If this were true, it really would seem to follow that Burt's data were critical, and that proof of their fraudulence was tantamount to proof that the entire hereditarian case was based on fraud. I believe that this conclusion can no more be justified than can Jensen's and Eysenck's claim that the removal of Burt's data made no difference to estimates of the heritability of IQ. At first glance, Kamin's and Taylor's analyses may seem remarkably convincing. But first glances often deceive. The critical question is whether they have been selective in their reporting, or biased in their criteria for including a pair of twins in one or other of their sub-samples. For example, in response to Kamin's criticisms, Shields (1978) himself reported a re-analysis of his own data: the correlation for the 16 pairs selected as experiencing the most similar environments was 0.87, that for the 12 pairs selected as experiencing the least similar environments was 0.84. Bouchard (1983) noted that the criteria which Taylor used to define five pairs of Shields' twins with quite different upbringings applied equally to a further six pairs of twins from the other two studies. For this larger sample of eleven pairs, the IQ correlation was 0.67.

One need not stop there. Kamin's seven pairs of twins from Shields' study with highly correlated environments, whose IQ scores correlated at the impossibly high figure of 0.99, can only have been selected largely on the basis of the close similarity in their IQ scores! Using even stricter criteria than Kamin casually reports, one can find another seven pairs, all brought up by relatives, all living close to one another, and attending the same school for several years. The correlation in their IQ scores is only 0.56. Bouchard (1983) has reported an ingenious check on the validity of Taylor's analyses of the effect of age of separation, reunion in childhood, related families, and similarity of social circumstance. Those analyses were based on the results of only one of the IQ tests given to the twins in the Newman *et al.* and Juel-Nielsen studies. When Bouchard repeated the analyses using the results of the second IQ test given, he found an entirely different pattern of results: age of separation *did* now seem to affect the resemblance in IQ scores, but the other three variables now did not.

One of the few conclusions that we can confidently draw from all this is that if a critic is determined to find support for a given position, a sufficiently lengthy trawl through a limited data set will surely unearth some aspects of the data that can be used to argue for that position. This usually reflects more on the nature of the critic's preconceptions than on the data themselves. The one sure way of guaranteeing the truth of the adage that you can prove anything with statistics is when you know in advance what you wish to prove; you allow yourself to divide, on the basis of any criteria you choose, already small samples into even smaller sub-groups, on which you perform as many statistical tests as you wish; and finally, you ignore the tests which do not give the results you want and report only those which support your position. Conclusions reached in this way are not worthy of anyone's serious attention.

Where, however, does this leave our original question? How important were Burt's MZa twin data? The answer is surely that they were never critical. There was other evidence, from other studies of separated twins, to suggest that IQ had significant heritability. Indeed, there was good evidence for such a conclusion from quite different kinds of study. Two classic American adoption studies had shown that the resemblance in IQ between adopted children and their adoptive parents was signficantly less than the resemblance between genetically related children and parents in ordinary 'biological' families (Burks 1928; Leahy 1935). A later study had shown that adopted children resembled in IQ their 'biological' mother, with whom they had never lived, more closely than their adoptive mother, with whom they had spent all their life (Skodak and Skeels 1949).

But this does not mean that Eysenck and Jensen were justified in arguing that the loss of Burt's data had no impact on the hereditarian case. Their loss was important because, on the face of it, his data were less problematic than those of other investigators, less confounded by other factors, less open to alternative interpretations. What this means, in other words, is that Kamin's and Taylor's criticisms, although in the end unconvincing, cannot simply be ignored. The studies they criticized were far from flawless. By their very nature, natural experiments are unlikely to be flawless. Many of the data in the social sciences come from natural experiments; their results are usually equivocal, open to more than one interpretation. Those on the heritability of IQ, whether they come from studies of separated MZ twins or other kinds of adoption study, are and were no exception—and one reason for being suspicious of Burt's studies is precisely that they seemed to be virtually immune to this sort of problem.

The position today

As I have already argued, the consensus view is that IQ has significant, or even substantial, heritability. But it is worth recollecting that this was the consensus view of most IQ testers long before any serious evidence on the question became available. Burt (1909), Spearman (1914), and Terman (1916), for example, all argued that intelligence was largely inherited in the absence of any evidence from adoption studies, let alone a study of separated MZ twins. If Kamin's or Taylor's conclusions are suspect because it seems probable that they knew what conclusion they wanted to draw before they examined the evidence, may not the same be said of the conclusions of hereditarians?

There can be only one answer: of course it can. Both hereditarians' and environmentalists' conclusions should be suspect. But that is only to say that *no* conclusion should be taken on trust here; *all* should be critically examined. In a review of Kamin's 1974 book, I wrote:

By his attention to detail, his ferreting out of inconsistencies, and his insistence on returning to the primary sources, Kamin gives us a salutary reminder of what scholarship is all about. No science can fail to benefit from this sort of meticulous analysis, and Kamin has put all future workers in this area profoundly in his debt . . . I think it undeniable that Kamin has shown that both in quality and quantity the evidence for the heritability of IQ is very much less than the consensus view has suggested. The data are sparse rather than plentiful, and at best persuasive rather than decisive. (Mackintosh 1975, pp. 684–5)

That still seems to me a fair judgment. Kamin had done his best utterly to demolish the evidence for the heritability of IQ. In the judgment of most serious reviewers, he did not succeed. But he did succeed in disposing of some of that evidence, and in casting doubts on some other parts. Instead of resting on their laurels, behaviour geneticists had to sharpen their arguments, design new, more careful studies, obtain fresh evidence. They have done so. The Minnesota twin study, still in progress, has so far reported IQ data on some 50 pairs of separated MZ twins. On different tests, their scores correlate between 0.69 and 0.78 (Bouchard *et al.* 1990). Although the twins were not all separated before the age of six months, their average age of separation was five months, and there is no correlation between age of separation or total amount of time they have been in contact with one another and the similarity in their IQ scores. Their adoptive homes do resemble one another in certain respects, but the resemblances are not close and do not appear to be in features that actually have any great effect on their IQ scores. These results suggest a figure for the heritability of IQ similar to that espoused by Woodworth and Eysenck.

Interestingly enough, however, this is probably no longer the consensus view. By 1980, several behaviour geneticists were suggesting a figure nearer 0.50, or within the range 0.40 to 0.70, for the heritability of IQ (e.g. Henderson 1982; Scarr and Carter-Saltzman 1982), and a more recent, reasonably sophisticated attempt to model a wide variety of kinship data also came up with a value of 0.50 (Chipuer *et al.* 1990).

These estimates are mostly dependent on the results of several American adoption studies (e.g. Horn *et al.* 1979; Scarr and Weinberg 1983; Phillips and Fulker 1989), rather than on the separated MZ twin data, which would still be consistent with a rather higher estimate. These adoption studies, which include families with both adopted and their own natural children, have generally found that correlations between biologically related people (brothers and sisters, parents and children), are higher than those between unrelated people (two adopted children living in the same family; an adopted child and a natural child of the adoptive parents; adopted children and their adoptive parents). In most cases, this is still true even when the 'biological' correlation is between adopted children and their biological mothers who gave them up for adoption. This pattern of difference implies that IQ has significant heritability, but some of the differences are quite small and the correlations vary from one study to another. Hence the overall estimate of the heritability of IQ is smaller than it once was.

No one is denying that IQ is affected by environmental circumstances: an estimate of only 50 per cent heritability, after all, leaves scope for substantial environmental effects. One intriguing observation of one of the more recent adoption studies, however, is that resemblances between unrelated adopted children living in the same family, or between such children and their adoptive parents, appear to *decrease* as the children grow older (Loehlin *et al*. 1989). The implication, which is consistent with the results of other studies of older adopted children (e.g. Scarr and Weinberg 1983; Teasdale and Owen 1984), is that the effects on IQ of the common (adoptive) family environment decrease as children grow older—and presumably spend more time outside the home, at school, or with their own friends. Critics of IQ tests have often argued that differences in test scores are largely a consequence of the huge differences in wealth, social class, and educational opportunity which an unjust society allows to exist. The intriguing possibility suggested by these newer findings is that many of the environmental sources of variation in IQ occur *within* families, rather than between one family and another, in other words that they are not a matter of these socio-economic factors at all.

No doubt, future research will modify some of these conclusions. Some should certainly be regarded as more tentative than others. Not all have been consistently supported by all the available evidence. That evidence is still of variable quality, and all of it is the product of imperfect natural experiments. My own view, not shared by all behaviour geneticists, is that there is little point in getting too exercised over the question of whether the heritability of IQ is really 0.70, 0.50, or even 0.30. There is, after all, no such thing as the 'true' heritability of IQ—that is to say, true for all times and places. Heritability is a statistic true for a given population at a given time, and will vary as circumstances change. Increases in social equality, or equality of educational opportunity, will probably eliminate some environmentally caused variation in IQ scores. But that necessarily implies that such changes will *increase* the heritability of IQ—i.e. increase the proportion of variance in IQ attributable to non-environmental differences. I do not believe that we know what the 'true' heritability of IQ is in North America or Western Europe today. It is quite certainly greater than zero, most probably greater than 0.30. That conclusion is most unlikely to be overturned by new experimental evidence, or the re-analysis of old. It is the most plausible interpretation of a wide range of data, of incomparably better quality, more carefully reported and analysed, than any data published by Burt himself. It is not an idiosyncratic hypothesis, held only by a small number of reactionary zealots. It is not just Burt's theory, and it no longer in any way depends on Burt's data (even if it once did). In that sense, Burt's fabrications—if that is what they were—did *not* seriously mislead the long-term course of scientific enquiry.

Epilogue

But were his data fabricated? Was Burt a fraud? It is tempting to say that readers of this book should make up their own minds. No doubt they will. But perhaps I too should come off the fence.

My own judgment is that the cumulative weight of the evidence makes it difficult to maintain Burt's innocence. Certainly, he is not innocent on all the charges brought against him. Some of Burt's more enthusiastic defenders, such as Fletcher (1991), have claimed that there is nothing wrong with his later empirical studies other than occasional carelessness, more or less excusable in a man in his eighties; they have insisted that anyone who suggests otherwise is applying later standards of critical rigour and methodological purity to his work: judged by the standards of his own time, they argue, Burt's studies are meticulous in their attention to detail, and where his data are less than perfectly reliable, he is the first to acknowledge their imperfections. I do not believe that this defence stands up to serious scrutiny. Quite apart from anything else, I think that it does a disservice to an earlier generation of social scientists. As Jensen (1974) has pointed out, although Burt's theoretical analyses were often much more sophisticated than those of his contemporaries, his failure to provide a proper account of his data, whether or not that failure amounts to fraud, has rendered his data, unlike theirs, useless to posterity. One has only to read the reports of their studies by Burks (1928), Leahy (1935), Newman *et al.* (1937), and Shields (1962) to see that, by comparison, Burt's presentation of his data is nearly always casual, often (one suspects) deliberately vague and imprecise, and sometimes dishonest.

Does this matter? In one sense, of course it does. These are not trivial charges. But do they constitute evidence of fabrication? Not necessarily. Thus any reasonable reading of Conway (1958) or Burt (1966) implies that the data for more than 50 per cent of his separated MZ twins were collected after 1955. We can be perfectly confident that this is not true. Indeed, I believe that the only sensible position for the defence to adopt here is that essentially all of Burt's MZa twin data were collected before 1945. But the fact that Burt implied otherwise does not prove that he simply invented these additional twins. Their data may have been mislaid and Burt may have been reluctant to admit this, as Banks (1983) and Joynson (1989) suggest, partly out of deference to the sensibilities of a secretary who was bad at filing and who would have felt responsible for the temporary disappearance.

But it is not always so easy to find such a charitable interpretation. Why, for example, was Burt sometimes so vague and obscure about the details of his samples and test procedures? What is the explanation of the fluctuations in the sample size for his other kinship correlations published in 1955 and 1966? Why does the 1969 paper on educational standards give no information whatsoever about the size of his successive samples? Why are we left to guess what tests were used for the later samples? Why does the 1961 paper on social mobility give no information on the actual size of the samples of adults and children that appear to a base of 1000 in Tables I and II (see Tables 4.1 and 4.2) of that paper? Why does Burt say that the number of adults in social class I actually studied was nearer 120 than 3? Was it 120, or 62, or 500? Why can he not say? And why does he not state that the source used in this paper for the estimates of the distribution of occupational classes in the population as a whole was Spielman and Burt (1926)?

It is difficult to resist the suspicion that Burt has something to hide, and that in some cases at least this ambiguity was deliberately designed to conceal serious inadequacy in the data—or worse. In the 1969 paper, it seems clear enough that Burt was concealing either that he had no data for his later samples, or that the data he had were based on new tests whose norms had never been compared with those of his pre-war tests, and whose results were therefore valueless. In his 1961 paper, he would have invited ridicule had he acknowledged that his estimates of the distribution of occupational classes were from Spielman and Burt (1926), who had acknowledged that their estimates were 'nothing more than the roughest approximation', which were based on an earlier census (presumably that of 1921) and on Charles Booth's London surveys at the turn of the century. It is equally clear that his assessments of adult intelligence in this paper were based on brief interviews, sometimes supplemented by 'camouflaged' tests. Does *this* matter? Did not Burt acknowledge that the assessments of adult intelligence were 'less thorough and reliable'?

This will not really do. The point, made above in another context, is that no trust can be placed in data such as these obtained by someone who knows in advance what results he wants. Table I of Burt's 1961 paper purports to show the distribution of adult intelligence according to occupational class. The scores range from an IQ of less than 60 to one of over 140, and are grouped into ten bins, each spanning a range of ten IQ points (i.e. 50–60, 60–70, etc.). How could a brief interview have been sufficient to decide into which of these ten bins someone should be assigned? But the scoring was, in fact, even more precise than this: the mean IQ of each occupational class cannot be reconstructed by assuming that everyone in bin 80–90, say, was assigned an IQ of 85. So we are asked to believe that a brief interview was sufficient to decide whether someone's IQ was 89 or 95 or 101. This is simply not realistic.* But Burt's entire argument in this paper required him to establish certain quite specific facts about the distribution of IQ by occupational class. Even if he did have some rough and ready assessment of the intelligence of some unspecified number of adults, there can be no reason to believe (to take examples at random) that exactly 75 per cent of adults in class IV, but less than 8 per cent of those in class VI, had IQs above the population mean of 100, or that only 7 per cent of adults in class III, but 26 per cent of their children had IQs below 100. It seems more than likely that some of these assessments were 'adjusted', i.e. moved from one side of a borderline to another, to give the answers he wanted.

Is this an unreasonable suspicion? I think not. Is it really plausible to suppose that the perfect fit between data and theory which Burt achieved here could have come about without assistance of this nature? Moreover, it is here, so it seems to me, that the cumulative weight of the evidence begins to tell. Eysenck provides a number of examples to show that Burt could behave dishonestly in order to advance his own cause, and could even invent theses apparently documenting points he wished to make. As Blinkhorn shows, there really must be some suspicion attached to the dates which Burt claimed for the writing and publication of the critical appendix on simple summation which eventually appeared in later editions of *Mental and scholastic tests*. Taking factors such as this into consideration, I do not believe that the defence can expect anyone to trust the numbers presented in Burt's 1961 paper. But were they actually fabricated? Burt, the defence can argue, would surely not be the first, or last, scientist to nudge his data in the direction he wished to see them go; but if that is all that can be proved against him, the defence may rest content: this is not what the prosecution meant by fabrication.

* The most detailed account of the procedures Burt and his assistants adopted for the assessment of parents' intelligence quite explicitly states that their assessments were much cruder than this. In the *Study of vocational guidance* (1926), undertaken under Burt's direction, Ramsey states:

The homes of the children tested were specially visited; and, in accordance with the general scheme drawn up by Dr Burt, the following particulars, where possible, were obtained . . . An attempt was made to grade the intelligence of the mother—the parent most often interviewed. Only a rough estimate could be made during an interview lasting from 20 to 40 minutes; and of course this estimate was entirely a personal one based upon the judgment of one investigator only. Notes were made immediately on leaving the house, and the mother described as (1) Very Intelligent (A+); (2) Intelligent (A); (3) Moderately Intelligent (B); or (4) Unintelligent (C). (Ramsey 1926, pp. 72, 75)

In a footnote, Ramsey states that these four grades correspond to the average IQ scores supposedly required for various occupational classes, such that A+ corresponds to classes I, II and III, with an average IQ of 124; A corresponds to class IV, IQ 107; B corresponds to class V, IQ 92; C corresponds to class VI, IQ 79.

I do not believe it possible to draw such a hard and fast distinction between adjustment and fabrication of data. One activity merges imperceptibly into the other, and inventing numbers that purport to represent data, when you have collected no data at all, is not the only way of committing scientific fraud. That is illustrated clearly enough by one of the less ambiguous problem cases. The 1969 educational standards paper purported to demonstrate a one-point decline in the intelligence of London schoolchildren between 1914 and 1965. Proper data, we now know, would have been more likely to show something like a ten-point increase. It seems probable that Burt either simply fabricated the numbers for his later surveys, or relied on new tests whose norms he had not properly compared with those of his earlier tests, and whose results were therefore quite invalid. The point is that in either case he committed scientific fraud, and in the latter case he attempted to cover this up by falsely claiming that he had collected sufficient data in about 1960 to allow him to check the norms for the 1962 edition of *Mental and scholastic tests*.

The problem with the data in the 1969 paper, as I have argued above, is not as Hearnshaw (1979) suspected that Burt could not have collected them; it is that they could not possibly be true. Ironically enough, I believe, much the same is true of the other cases of suspected fraud: that is to say, the most suspicious features of Burt's later claims and papers are usually *not* those actually identified by his initial critics. Thus Dorfman's (1978) main evidence of fraud in Burt's 1961 paper on intelligence and social mobility was that the IQ scores were too perfectly normal, and that the proportions of adults in each social class perfectly matched Spielman and Burt's (1926) estimates. But Burt said that he had weighted his class frequencies to match *estimated* frequencies in the population, and although the use of the Spielman and Burt estimates was absurd, it was not fraudulent. And Dorfman's main proof that Burt's data were too perfectly normal rested on the erroneous assumption that his samples were 40000 strong. The critical problem with these IQ data is not their perfect normality, as originally argued by Dorman, it is the departures from normality noted by Rubin (1979): those departures are not random, but show every sign of fabrication.

Finally, the initial accusations of fraud, those concerning the MZ twin data, were based on the invariant correlations and the missing assistants. But neither is decisive. The invariant correlations are evidence of extraordinary muddle and carelessness in the 1966 paper; the missing assistants prove only that the data were not collected after 1955. Hearnshaw's evidence from the diaries and correspondence proves little more. We can be reasonably confident, indeed, that essentially all the data reported in 1955 and 1966 had been collected before 1945, most of them before 1932. So it is bizarre that Burt should have kept silent about them for so long, and even more bizarre that he should have implied in 1931 that no one had used the

148   *Does it matter? The scientific and political impact of Burt's work*

London School system as a source of twin data. But the most serious problem with the finally published twin data is the discrepancy between the numbers Burt sent to Jencks in 1969 and Burt's (1966) and Conway's (1958) earlier statements, coupled with Burt's diary entry in which he talked of 'calculating twin data for Jencks'. It is this that suggests that the correlations reported by Burt and Conway were not based on real numbers, and that Burt had been forced to work out some new numbers in 1969 in order to provide the scores which Jencks had requested. And a final cause for suspicion, as Jensen notes in his chapter, is Burt's claim to Sandra Scarr in 1971 that he had recently obtained data on three new pairs of twins.

None of these cases, taken alone, is necessarily decisive—sufficient to establish guilt, in a court of law, beyond reasonable doubt. Of each, taken in isolation, the defence could reasonably say that there must be an alternative explanation—even if we do not know what it is. And since Burt is no longer alive to provide that defence, the only reasonable verdict must be: not proven. But we are not trying a case in a court of law. We are simply attempting to arrive at the most plausible explanation of the available evidence. And when all these cases are considered together, a defence which can only insist that there *must* be some other, more innocent explanation begins to lose its force. One can appeal to an unknown, even somewhat implausible, explanation once. To have to do so two or three times makes one's case rather less persuasive. We may say that it is implausible to suppose that a man of Burt's stature and reputation should have committed fraud. No doubt it is. No doubt it is implausible to suppose that Newton did. But if the defence is equally implausible, the case for the prosecution starts to look that much more convincing. We know that Burt could be devious and dishonest in small things, and was sometimes determined to win at any cost. We know that he had a relatively cavalier attitude to the reporting of empirical data. We know that he was prepared to conceal things about his data that would have made those data very much less persuasive, and we can be confident that he was sometimes prepared to adjust his data, and at other times make false claims about them, in order to make them appear more convincing. On balance, I believe that the evidence makes it more probable than not that some of the data he reported existed only in his imagination, in other words that he fabricated them.

References

Banks, C. (1983). Professor Sir Cyril Burt: selected reminiscences. *Journal of the Association of Educational Psychologists*, **6**, 21–42.
Bouchard, T. J. Jr. (1983). Do environmental similarities explain the similarity in intelligence of identical twins reared apart? *Intelligence*, **7**, 190–1.

Bouchard, T. J. Jr. and Segal, N. L. (1985). Environment and IQ. In *Handbook of intelligence: theories, measurements, and applications* (ed. B. B. Wolman), pp. 391–464. Wiley, New York.

Bouchard, T. J. Jr., Lykken, D. T., McGue, M., Segal, N. L., and Tellegen, A. (1990). Sources of human psychological differences: the Minnesota study of twins reared apart. *Science*, **250**, 223–8.

Burks, B. S. (1928). The relative influence of nature and nurture upon mental development: a comparative study of foster parent–foster child resemblance and true parent–true child resemblance. *Yearbook of the National Society for the Study of Education*, **27**, 219–316.

Burt, C. L. (1909). Experimental tests of general intelligence. *British Journal of Psychology*, **3**, 94–177.

Burt, C. L. (1961). Intelligence and social mobility. *British Journal of Statistical Psychology*, **14**, 3–24.

Burt, C. L. (1966). The genetic determination of differences in intelligence: a study of monozygotic twins reared together and apart. *British Journal of Psychology*, **57**, 137–53.

Burt, C. L. (1969). Intelligence and heredity: some common misconceptions. *Irish Journal of Education*, **3**, 75–94.

Chipuer, H. M., Rovine, M. J., and Plomin, R. (1990). LISREL modeling: genetic and environmental influences on IQ revisited. *Intelligence*, **14**, 11–29.

Conway, J. (1958). The inheritance of intelligence and its social implications. *British Journal of Statistical Psychology*, **11**, 171–90.

Dorfman, D. (1978). The Cyril Burt question: new findings. *Science*, **201**, 1177–86.

Dorfman, D. (1979). Correspondence. *Science*, **204**, 246–54.

Eysenck, H. J. (1973). *The inequality of man*. Temple Smith, London.

Farber, S. L. (1981). *Identical twins reared apart: a reanalysis*. Basic Books, New York.

Fletcher, R. (1991). *Science, ideology, and the media: the Cyril Burt scandal*. Transaction Publishers, New Brunswick, NJ.

Flynn, J. R. (1984). The mean IQ of Americans: massive gains 1932–78. *Psychological Bulletin*, **95**, 29–51.

Flynn, J. R. (1987). Massive IQ gains in 14 nations: what IQ tests really measure. *Psychological Bulletin*, **101**, 171–91.

Gillie, O. (1976). Crucial data was faked by eminent psychologist. *The Sunday Times*, 24 October. London.

Goldthorpe, J. H. and Hope, K. (1974). *The social grading of occupations*. Clarendon Press, Oxford.

Hearnshaw, L. S. (1979). *Cyril Burt: psychologist*. Hodder and Stoughton, London.

Henderson, N. D. (1982). Human behavior genetics. *Annual Review of Psychology*, **33**, 403–40.

Horn, J. M., Loehlin, J. C., and Willerman, L. (1979). Intellectual resemblance among adoptive and biological relatives: the Texas adoption project. *Behavior Genetics*, **9**, 177–207.

Jencks, C. (1972). *Inequality: a reassessment of the effect of family and schooling in America*. Basic Books, New York.

Jensen, A. R. (1969). How much can we boost IQ and scholastic achievement? *Harvard Educational Review*, **39**, 1–123.

Jensen, A. R. (1972). *Genetics and education*. Methuen, London.

Jensen, A. R. (1974). Kinship correlations reported by Sir Cyril Burt. *Behavior Genetics*, **4**, 1–28.

Joynson, R. B. (1989). *The Burt affair*. Routledge, London.

Juel-Nielsen, K. (1965). Individual and environment: a psychiatric-psychological investigation of MZ twins reared apart. *Acta Psychiatrica Scandinavica* (Suppl. 183). Munksgaard, Copenhagen.

Kamin, L. J. (1974). *The science and politics of IQ*. Erlbaum, Potomac, MD.

Kamin, L. J. (1981). *Intelligence: the battle for the mind* (H. J. Eysenck versus Leon Kamin). Macmillan, London.

Leahy, A. (1935). Nature–nurture and intelligence. *Genetic Psychology Monographs*, **17**, 236–308.

Loehlin, J. C., Horn, J. M., and Willerman, L. (1989). Modeling IQ change: evidence from the Texas adoption project. *Child Development*, **60**, 993–1004.

Mackintosh, N. J. (1975). Review of Kamin (1974). *Quarterly Journal of Experimental Psychology*, **27**, 672–86.

Mascie-Taylor, C. G. N. and Gibson, J. B. (1978). Social mobility and IQ components. *Journal of Biosocial Science*, **10**, 263–76.

Newman, H. H., Freeman, F. N., and Holzinger, K. H. (1937). *Twins: a study of heredity and environment*. University of Chicago Press.

Phillips, K. and Fulker, D. W. (1989). Quantitative genetic analysis of longitudinal trends in adoption designs with application to IQ in the Colorado adoption project. *Behavior Genetics*, **19**, 621–58.

Ramsey, L. (1926). Home conditions. In *A study in vocational guidance*, Medical Research Council, Industrial Fatigue Research Board, Report No. 33, pp. 72–8. HMSO, London.

Rubin, D. B. (1979). Correspondence. *Science*, **2**, 245–6.

Rushton, J. P. (1994). Victim of scientific hoax. *Society*, **31**, 40–4.

Rutter, M. and Madge, N. (1976). *Cycles of disadvantage*. Heinemann, London.

Scarr, S. and Carter-Salzman, L. (1982). Genetics and intelligence. In *Handbook of human intelligence* (ed. R. J. Sternberg), pp. 792–896. Cambridge University Press.

Scarr, S. and Weinberg, R. A. (1983). The Minesota adoption studies: genetic differences and malleability. *Child Development*, **54**, 260–7.

Shields, J. (1962). *Monozygotic twins brought up apart and brought up together*. Oxford University Press, London.

Shields, J. (1978). MZ twins: their use and abuse. In *Twin research: psychology and methodology* (ed. W. Nance). Liss, New York.

Skodak, M. and Skeels, M. H. (1949). A final follow-up study of one hundred adopted children. *Journal of Genetic Psychology*, **75**, 85–125.

Snyderman, M. and Rothman, S. (1988). *The IQ controversy: the media and public policy*. Transaction Publishers, New Brunswick, NJ.

Spearman, C. (1914). The heredity of abilities. *Eugenics Review*, **6**, 219–37.

Spielman, W. and Burt, C. L. (1926). The estimation of intelligence in vocational guidance. In *A study in vocational guidance*, Medical Research Council, Industrial Fatigue Research Board, Report No. 33, pp. 12–17. HMSO, London.

Taylor, H. F. (1980). *The IQ game: a methodological inquiry into the heredity–environment controversy*. Rutgers University Press, New Brunswick, NJ.

Teasdale, T. W. and Owen, O. R. (1984). Heredity and familial environment in intelligence and educational level—a sibling study. *Nature*, **309**, 620–2.

Terman, L. M. (1916). *The measurement of intelligence*. Houghton Mifflin, Boston, MA.

Van Court, M. and Bean, F. D. (1985). Intelligence and fertility in the United States: 1912–1982. *Intelligence*, **9**, 23–32.

Waller, J. H. (1971). Achievement and social mobility: relationships among IQ score, education and occupation in two generations. *Social Biology*, **18**, 252–9.

White, K. B. (1982). The relation between socioeconomic status and academic achievement. *Psychological Bulletin*, **81**, 461–81.

Index